# BECOMING GOOD ANCESTORS

# BECOMING
# GOOD
# ANCESTORS

## HOW WE BALANCE NATURE,
## COMMUNITY, AND TECHNOLOGY

David Ehrenfeld

OXFORD
UNIVERSITY PRESS
2009

# OXFORD
## UNIVERSITY PRESS

Oxford University Press, Inc., publishes works that further
Oxford University's objective of excellence
in research, scholarship, and education.

Oxford    New York
Auckland    Cape Town    Dar es Salaam    Hong Kong    Karachi
Kuala Lumpur    Madrid    Melbourne    Mexico City    Nairobi
New Delhi    Shanghai    Taipei    Toronto

With offices in
Argentina    Austria    Brazil    Chile    Czech Republic    France    Greece
Guatemala    Hungary    Italy    Japan    Poland    Portugal    Singapore
South Korea    Switzerland    Thailand    Turkey    Ukraine    Vietnam

Published by Oxford University Press, Inc.
198 Madison Avenue, New York, NY 10016

www.oup.com

Oxford is a registered trademark of Oxford University Press

Library of Congress Cataloging-in-Publication Data
Ehrenfeld, David.
Becoming good ancestors : how we balance nature, community,
and technology / David Ehrenfeld.
p.   cm.
Rev. ed. of: Swimming lessons / David Ehrenfeld. 2002.
Includes bibliographical references and index.
ISBN 978-0-19-537378-3 (pbk.)
1. Human ecology.    2. Nature—Effect of human beings on.
3. Environmental degradation.    4. Civilization, Modern—21st century.
5. Technology—Social aspects.    I. Ehrenfeld, David. Swimming lessons.    II. Title.
GF49.E47 2009    304.2—dc22    2008007199

This book includes material previously published in the author's
*Swimming Lessons* (Oxford University Press, 2002).

Printed in the United States of America
on acid-free, recycled paper

*For my grandchildren*

# CONTENTS

# PREFACE

It was a fine, late summer day, and my wife and I were sitting in a pontoon boat moving upstream on the Hackensack River through the marshes of the Jersey Meadowlands. Fish were jumping. Manhattan was only five miles away to the east, across another river, the Hudson. The purpose of the trip was to see birds. We had passed a multitude of them, including snowy and great egrets, great blue herons, double-crested cormorants, greater and lesser yellowlegs, assorted seagulls, northern harriers, and a single Caspian tern, with its bright red beak. As our boat glided under a highway bridge, someone shouted, "Look!" There, sitting on a concrete bridge support not far above our heads was a large, female peregrine falcon, watching us calmly. Nearby, on other beams under the bridge, were two more peregrines, a male and another female.

Why begin my preface with this anecdote? Because a half-century earlier, the river was filthy, there were almost no fish in it, the only birds were seagulls feeding on garbage and hawks eating rats at the many dumps near the river, and the peregrine falcon was well on its way to what seemed certain extinction. Then, a fortuitous combination of events and people turned the situation around. The U.S. Clean Water Act eliminated unregulated garbage dumping. Andy Willner, of the American Littoral Society, started the New York/New Jersey Baykeeper;

and Bill Sheehan gave up his safe job as a cab driver and dispatcher to start the Hackensack Riverkeeper. These organizations mobilized concerned citizens throughout the regional watershed and became powerful forces to ensure that conservation gains were preserved. Meanwhile, Derek Ratcliffe, chief scientist of Britain's Nature Conservancy Council, discovered that DDT was the cause of the decline of the peregrines and other hawks and eagles, ultimately leading to widespread banning of the pesticide. And in the United States, Tom Cade, of Cornell, began an emergency program to release captive-reared peregrines into the wild. In 1974, when he released his first falcons, there were fewer than forty pairs nesting in the lower forty-eight states. By 1999, the number of nesting peregrines had grown to more than two thousand, and the bird was removed from the endangered species list. Nature is resilient if we give it a chance.

Victories like these are a true cause for celebration. But they are all too infrequent, occurring only when highly specific, defined problems have equally specific and defined solutions—regulate garbage dumps and other point sources of pollution; form effective, local nonprofit organizations to monitor local habitats and lobby for their protection; ban the use of toxic chemicals that are persistent in the environment; and breed replacements for the animals that the chemicals affected. Unfortunately, most of our problems these days are not so simple to understand or straightforward to solve, and they are growing in severity and number.

The complex interactions among society, technology, and the natural world in the twenty-first century may seem beyond our comprehension, yet there are underlying, destructive patterns that we can grasp. Some of them are: retreat from reality, alienation from nature, unthinking acceptance of new technology and rejection of the old, loss of our ability to discriminate between events we can control and those we cannot, denial of noneconomic values, and the disempowerment of local communities.

*Becoming Good Ancestors* is an examination of these destructive patterns through down-to-earth examples, and a discussion of how we can use what we learn to move ourselves and our society toward a more stable, less frantic, more responsible, and far more satisfying life. A life in which we are no longer compelled to damage ourselves and our environment. A life in which our parents and our children have a future. A life in which we hear more truth than lies. A life in which fewer species are endangered and more rivers run clean.

Such goals are not utopian pipe dreams. As the example of the Hackensack River and the peregrines shows, our society has an inherent sense of what is right and the creativity and persistence to make good things happen. It is now time to apply our intelligence guided by our moral judgment to the very largest difficulties we all face. We are in the eye of the future and should behave accordingly.

This book is an expanded, extensively revised, and updated version of my book *Swimming Lessons*, published in 2002. The chapters in *Swimming Lessons* first appeared as "Raritan Letters" in *Orion* magazine, and I thank *Orion* for giving me permission to reprint them. My editor at *Orion*, Aina Barten, worked with me on all of the Raritan Letters, providing her invaluable critical judgment of both the literary style and the content of what I had written. Emerson Blake, then the managing editor at *Orion*, was also immensely helpful and supportive.

My present editor at Oxford University Press, Peter Prescott, first suggested that I should turn the original book into something new, with greater cohesion and sharper focus—a book that would meet the real needs of the twenty-first century. And it was Peter who, with kindly but pointed criticism, prompted me to revisit every sentence and every claim in the older book, in a rewriting that took far longer than I had first anticipated. I thank him for his wisdom, his confidence, and his endurance.

Many people have helped and inspired me in my work. I am especially grateful to my friends Wendell Berry, David Orr, and Wes Jackson, undergraduates too numerous to name, and my present graduate students Kristi MacDonald-Beyers and Joe Paulin. I also thank professor Kristen Conway-Gómez and the other faculty and students in the First Year Experience course at California State Polytechnic University, Pomona, for their insightful criticisms and suggestions.

As the readers will discover, my ideas are inseparable from my life and experiences, and this means inseparable from my children: Kate and her husband Dan, Jane and her husband Michael, Jon, and Sam. They are woven into my stories as they are woven into my existence, both of which are enriched beyond my power to thank or describe. And as they grow older, I have found their comments about what I

"Keeping Track of Our Losses," the second section, is a response to a world changing so rapidly that it is easy to lose sight of the things that have worked well in the past yet have been swept aside in the tsunami of what is called progress. Of course we can say good riddance to many of the evils in the lives people lived one hundred years ago, but can we afford to forget all the social inventions and practical ways of living that have helped us survive up to this point? Here I describe precious features of our former life that are disappearing—stability in the workplace; the ability to take time to make important decisions; the joy of maintaining deep relationships with people who are physically close enough to be seen, heard, and touched; the pleasure of reading the great and lesser works of our literature and being able to write in a way that does credit to our remarkable language. And I cast light on the dark mechanisms that help us forget them—ready adaptation to conditions that we ought to resist; galloping obsolescence that destroys perfectly useful inventions; and the increasingly rapid development of electronic substitutes for human beings. If we are aware of what we are losing and why we are losing it, we have a better chance to halt and reverse the cultural erosion that is taking place around us.

Part 3, "Toward a Sustainable Economics," is an account of what happens to us and to our environment when we translate all of our values into economic terms and subordinate all of our decisions to techno-economic imperatives and the ethics of globalization and growth. This is a wide-ranging section of the book—it goes from tuna-fishing fleets to Chinese turtle and tiger farms to corporations that live forever and seem beyond the law; from Baffin Island to Lenapehoking (now New Jersey) to the Venezuelan rain forest. Economics has stopped being practical, if it ever was, and has lost touch with the world we live in. When economic beliefs and practices are dissociated from their effects and consequences, and the feedback that fosters positive change disappears, bad things happen. If we stop looking at everything solely through an economic lens, it is much easier to see what we have to do.

"Relating to Nature in a Human-Dominated World," part 4, has more chapters than any other part of the book, yet it is least in need of a bookmap. I think that the unifying themes are simple and evident: First, we need regular, frequent, informal contact with nature to keep ourselves and our world healthy, but modern life increasingly walls us (especially our children) off from the living environment. Second, despite the amazing discoveries of our science and the more amazing claims of its publicists, many of nature's workings remain a mystery to us. Third, there are many ways to reconnect with nature. Fourth, nature can and usually does regenerate by itself if you let it; our substitutes for nature don't. And fifth, nature presents a parade of fascinating surprises. A few of the surprises that I discuss are: thousand-year-old black spruce trees in Quebec that are little more than a foot tall; a six-hundred-pound female Alaskan grizzly bear ordering her two cubs to stay away from a terrified hiker who has fallen down a few feet from her path; huge Louisiana alligators that, in the nineteenth century, refrained from attacking the little boys who often swam next to them and pelted them with mud for fun; and English lime (linden) trees that won't spread to apparently suitable parts of the countryside adjacent to their present stands if these adjacent areas are outside the boundaries of what once were the Neolithic forestlands where lime trees used to grow.

The last section, "Restoring the Community," is neither a hearkening back to good old communities that never existed nor an anticipation of utopian communities that never will exist. Rather, it is a recognition (perhaps prediction would be a better word) that in a world constrained by growing shortages of energy and raw materials, and fractured by the breakdown of globalization, local communities will regain a good deal of their former importance. In other words, local communities will be coming back, like it or not, and it will be greatly to our advantage if we learn the elements that make them good places in which to live, places that reconcile human freedom and responsibility in harmony with the environment. We also need to learn how to avoid or ameliorate those elements that can make local communities terrible. This is not going to be an easy task, and it certainly will be done in many ways, according to circumstances. There is no single blueprint for community restoration. However, we can be aware of the place of traditions; we can try to identify the social mechanisms that stabilized communities in the past; we can learn how to make our schools and universities more responsive and useful to the neighborhoods around them; and we can discover how to raise children who will be assets to their communities. Throughout the section on "Restoring the Community," and in the sections before it, my arguments are tied to one central assumption: The way we behave and the values we espouse are at least as important for repairing the world as the plans we make for improving the future. Finally, in the epilogue, I explore those tenets of conservative and liberal philosophies that will have to be brought together if community, national, and planetary restoration is to succeed, and I look at some early, practical ways in which this is being accomplished.

There is a great deal of overlap among the five sections: for example, nature and ecology appear throughout the book. The themes of these sections are related; they are all part of a larger whole, my vision of what we have to know and have to do to make our way of life sustainable, deeply enjoyable, and fulfilling—a life that will make us good ancestors for our descendants.

The vertigo of the twentieth century needs no permission of yours or mine to continue. The tornado has not consulted any of us, and will not do so. This does not mean that we are helpless. It only means that our salvation lies in understanding our exact position, not in flattering ourselves that we have brought the whirlwind into being by ourselves, or that we can calm it with a wave of the hand.

Thomas Merton, "A Letter to Pablo Antonio Cuadra Concerning Giants"

# ONE

---

## IN SEARCH OF HONESTY

Never has failure been so ardently defended as though it were success.

John Ralston Saul, *Voltaire's Bastards*[1]

# PRETENDING

For as high as [the ostrich's] body is, yet if they thrust their head and neck once into any shrub or bush, and get it hidden, they think then they are safe enough, and that no man seeth them.

Pliny the Elder, *Natural History*[1]

My earliest memory—was I two? three?—is of a nurse or babysitter who was dressed in white and had a bad smell. When she came into my room, I would pretend to be asleep. She looked down at me (I could sense her presence), then, satisfied by my closed eyes and quiet, if shallow, breathing, would turn and leave the room. Soon her odor would go away, too, and I could breathe deeply again. This taught me a lesson of dubious value: when helpless in a situation, pretending can give you power. A small, weak child, I had controlled the movements of an enormous, smelly adult. It was some time later, I suppose, when I learned that pretending usually doesn't work. Although I can't recall the time or place, I know that on one grim day of disillusionment and reckoning I discovered that when I closed my eyes I didn't become invisible; I couldn't transport bullies to distant, foul places by imagining them there or alter the course of unwelcome events by pretending they were otherwise.

Healthy children come to know the difference between pretending that is relaxing, stabilizing, healing, necessary—we call it fantasy—and pretending that is dangerous. A hot fire burns, even if we pretend that it won't. In our personal lives most (but unfortunately not all) of us learn to instantly distinguish harmless from harmful pretending so that we do not pretend in a way that endangers our lives or physical well-being.[2] Strangely enough, this discrimination does not seem to apply to many common beliefs that we share within our society. For at least the past fifty years, probably longer, we have been working hard as a high-tech civilization to ignore the limits of safe pretending.[3]

To be fair to ourselves, our pretending is not exactly of the childhood fantasy variety—it is not derived from magical thinking, like the remarkable abilities of Harry Potter or the superhuman destructive powers granted to the players of electronic video games. Instead, our pretending springs from very real, solid beginnings, and this is what makes it so difficult to recognize and to control. When reality transitions seamlessly into pretending, and when both are certified by experts as true, how do we know where to draw the line?

In the next few pages, I give three case studies of this most dangerous kind of pretending, in which the initial reality camouflages the subsequent fantasy: genetic engineering in medicine and agriculture, the use of scientific models to predict future events, and the "Star Wars" research to develop ways of shooting down incoming enemy ballistic missiles. Although superficially dissimilar, these case studies have a number of features in common. There is an initial, well-intentioned, even benevolent goal that no right-minded person could reject—curing all genetic diseases and ending world hunger; predicting how long our beach stabilization measures or nuclear waste storage facilities will remain effective and safe; preventing the annihilation of cities by nuclear-tipped warheads loosed by a rogue power. In each case, there have been enough successes or partial successes to give some credence to our hopes. And in each it is finally a few of the experts themselves who say, "Enough! Our pretending is causing real damage, and it's time to stop."

The first case study is genetic engineering. In my last year of college, I was fortunate to have as one of my teachers a visiting professor from England, Francis Crick of Watson and Crick fame, who taught part of an upper-level biochemistry course on the structure and function of macromolecules, especially deoxyribonucleic acid, or DNA. Those were exciting days. Crick, a master teacher, explained his and Watson's model of DNA with brilliant clarity. The structure of DNA, the sequence of its four kinds of component molecules (bases), was like a book, in that it contained information that was then copied into a sister molecule, RNA. RNA, in turn, used this information to direct the synthesis of the many different proteins that formed or helped to synthesize all the molecular parts of the cell. This was the "central dogma" of molecular biology: DNA makes RNA makes protein. Genes are made of DNA, and the genetic "code" of the DNA molecule, it appeared, is the same for all organisms—plants, animals, microbes. Decipher the DNA code of one, several, or perhaps all the genes of a species, and the possibilities for manipulation and control of the growth, metabolism, form, and behavior of that organism are limitless. At least that's what we believed.

In the 1960s and '70s, the central dogma gave birth to the new field of genetic engineering. At the core of genetic engineering was transgenics, the semicontrolled transfer of genes from one individual to another, even across species boundaries. In the decades that followed, bacterial genes were moved into corn and soybeans; goats were engineered to express the silk-making genes of spiders; human genes were transferred to sheep, pigs, salmon, and other creatures. Genes have been deleted, added, and, to some extent, modified to achieve the desired results. The very nature of a species and the boundaries between species have come to be seen, at least to uncritical eyes, to be as malleable as soft clay. Genetic engineering, as portrayed by many of its practitioners and believed by much of the lay public, appears to have a limitless potential to modify humans and other life forms in desired directions—to cure most major diseases, to improve physique, to raise IQ, to overcome the limits of the most basic biological processes (such as photosynthesis), and to alter for the better any other biological attribute seen as undesirable. The student genetic engineering club at my college calls itself "Designer Genes."

Yet such claims fall far short of reality. Despite the enormous popular enthusiasm whipped up by the press and the financial markets, only a small proportion of the simplest possible genetic manipulations among the many that have been tried have worked at all. And many of these have turned out to be disappointing, dangerous, or both. The central dogma concerning the importance of DNA remains largely true, and its discovery is rightly celebrated as one of the greatest feats of twentieth-century science, but the precise control attributed to it is mostly imaginary. It has become increasingly apparent that DNA is only a part of the story—itself subject to other regulating and modifying influences in the cell, influences that we dimly understand. Rather than being at the top of a simple linear chain of command, the DNA of the gene should be seen as one piece of an interacting complex of regulatory systems and feedback loops, with no single element "running" the cell, let alone the entire organism.

Richard Lewontin, a senior molecular biologist at Harvard and a member of the U.S. National Academy of Sciences has said: "The bottom line is that life in all its manifestations is complex and messy and cannot be understood or influenced by concentrating attention on a particular molecule of rather restricted function."[4] Nowhere is this wisdom more studiously ignored than in the claims and promises of genetic engineers that our knowledge of the sequence of the human genome will lead to cures of most diseases. Lewontin has commented frankly about these claims:

> Unfortunately, it takes more than DNA to make a living organism. . . . According to the vision [of the genetic engineers], we will locate on the human chromosomes all the defective genes that plague us, and then from the sequence of the DNA we will deduce the causal story of the disease and generate a therapy. Indeed, a great many defective genes have already been roughly mapped onto chromosomes and, with the use of molecular techniques, a few have been very closely located and, for even fewer, some DNA sequence information has been obtained. But causal stories are lacking and therapies do not yet exist; nor is it clear, when actual cases are considered, how therapies will flow from a knowledge of DNA sequences.[5]

Why are the hopes of the public being raised so recklessly? There are many reasons, of course, including the unrealistic expectations that the public has of scientists. But I do not think it is cynical to cite another motive. I quote Lewontin again from the same reference: "No prominent molecular biologist of my acquaintance is without a financial stake in the biotechnology business." Sheldon Krimsky, a bioethicist at Tufts

University and former member of the National Institutes of Health Recombinant DNA Advisory Committee, has carefully documented the spectacular increase in recent years in the number of academic biotechnologists who are on the boards or scientific advisory committees of or own stock in the for-profit companies that are supporting their research. Krimsky asks:

> Can this pattern of behavior continue without any detriment to science? An editorial contributor to the *Journal of the American Medical Association* wrote: "When an investigator has a financial interest in or funding by a company with activities related to his or her research, the research is lower in quality, more likely to favor the sponsor's product, less likely to be published, and more likely to have delayed publication."[6]

The British molecular biologist and biophysicist Mae-Wan Ho has described the biological reasons that underlie the multiple failures and dangers of genetic engineering and related technologies such as cloning. In her very important and lucid book *Genetic Engineering: Dream or Nightmare?* Ho explains some of the biological challenges to genetic engineering. They include the instability of transgenes, overlapping genes, gene amplification and inactivation, environmentally induced changes in gene expression, inheritance that does not involve the DNA code, and even, possibly, the inheritance of acquired characteristics. She asks: "Would anyone think of investing in genetic-engineering biotechnology if they knew how fluid and adaptable genes and genomes are? The notion of an isolatable, constant gene that can be patented as an invention for all the marvellous things it can do is the greatest reductionist myth ever perpetrated."

In other words, the idea that these patented transgenic organisms (and there are now many) are genetically stable and capable of performing consistently as desired for long periods of time and through many generations, like a patented machine, is not biologically warranted. Ho also carefully evaluates the hazards of genetic engineering, particularly the increased risk of spreading antibiotic resistance among disease organisms and human and animal populations, and the risk that gene-transfer technologies will create new varieties of virulent pathogens. Citing the molecular geneticist Joe Cummins, Ho pointed out that "the powerful promoter from cauliflower mosaic virus, which is routinely used to drive gene expression in transgenic plants is closely related to the human hepatitis B virus and also has sequence homologies to human retroviruses such as the AIDS virus," and she warned that "two kinds of potential hazards can be envisioned: the reactivation of dormant viruses, and recombination between the CaMV promoter and other viruses . . . to generate new super-infectious viruses or viruses with a broadened host range."[7]

Genetic engineers have not reacted kindly to Ho's book. For example, in 2002, the genetic engineer M. Tepfer criticized Ho's fears of the CaMV promoter as "implausible"; yet in a 2004 paper, Tepfer and his colleagues admitted that "the multiple complexities and uncertainties involved in this work introduce an element of risk that cannot be ignored."[8]

In a well-documented case, the risks of creating genetically modified organisms were proven to be very real. At the Cooperative Research Center (CRC) for the Biological Control of Pest Animals and the John Curtin School of Medical Research at the Australian National University, both in Canberra, researchers produced a genetically altered mousepox virus designed to control rodent populations in Australia. To their horror, the recombinant virus wiped out all the experimental animals, even those that had been vaccinated against it. The leader of the research team, CRC director Bob Seamark, said: "This is the public's worst fears about GMOs come true. . . . We . . . had shown that a commonly used technique could overwhelm resistance and render vaccination useless." Anabelle Duncan, a former head of the United Nations inspectors in Iraq and chief of molecular scientists at Australia's Commonwealth Scientific and Industrial Research Organization (CSIRO) commented: "This shows that something we had thought was hard—increasing the pathogenicity of a virus—is easy." The only positive aspect of this terrifying episode is the honesty and public-spirited behavior of the scientists who conducted the research.[9]

The Australian researchers destroyed the deadly recombinant mousepox virus before it could escape to the outside environment. Unfortunately, because of the enormous number of recombinant organisms being tested by many laboratories in many venues, genetic hazards are not always discovered in time. In the so-called ProdiGene Incidents, corn engineered by a small Texas company to produce drugs for diabetes and diarrhea contaminated soybean fields in Iowa and Nebraska, as well as a Nebraska grain elevator. The U.S. Department of Agriculture ordered the 500,000 bushels of contaminated soy in the grain elevator destroyed, and 155 acres of contaminated fields in Iowa burned. The Grocery Manufacturers of America, speaking for an alarmed food industry, called for a moratorium on food crops genetically modified for drug production, pending strict government regulation.[10] But even the limited safety protocols that are in place are not necessarily followed.[11]

Some biotechnologists are scrupulously honest about the implications of their work. For example, in an article in the March 30, 2001, issue of *Science*, MIT's Rudolf Jaenisch and the Roslin Institute's Ian Wilmut (the cloner of Dolly) described the "drastic defects that occur during development" and the "high failure rates" that are part and parcel of the cloning of mammals; and they warn that the cloning of humans would be "dangerous and irresponsible," and will remain so for the "foreseeable future."[12] But these scientists may be exceptions. Commenting on biotechnology's frantic rush to create new organisms, Ho writes, "What the public is up against is a selective blindness to evidence among the genetic engineers and a single-minded commitment to look solely for the exploitable, which is the hallmark of bad science." In other words, most genetic engineers are pretending that their technology works as claimed, is stable, and is safe—that the euphoria of 1960 is still scientifically justified, in spite of what we have learned since then. Nor is the public entirely blameless in the matter. Especially in the United States, we have been all too willing to make the effortless choice of following the biotechnologists into uncharted territory, eyes and minds tight shut, delusions intact.

One of the most blatantly unrealistic forms of pretending is the faith that agricultural biotechnology, including genetic engineering to produce genetically modified organisms (GMOs), will end hunger. Peter Rosset, an expert on agricultural technology and codirector of Food First/The Institute for Food and Development Policy, in Oakland, California, has submitted this claim to critical analysis. There is no relationship, he finds, between the prevalence of hunger in a given country and the density of its population. "For every densely populated and hungry nation such as Bangladesh or Haiti, there is a sparsely populated and

hungry nation such as Brazil or Indonesia. In fact, per capita food production increases over the past four decades have far outstripped human
population growth. Thus, at a global scale, improved crop production
technology of any kind is unlikely to help."

This is true on average, but in some poor countries food production
is lagging seriously behind. Is biotechnology the answer in these countries? Rosset says no: "Third World food producers demonstrate lagging
productivity not because they lack 'miracle' seeds that contain their own
insecticide or tolerate massive doses of herbicide, but because they have
been displaced onto marginal, rain-fed lands, and face structures and
macroeconomic policies that are increasingly inimical to food production by small farmers."[13]

There is more, however, to the story than the many problems caused
by insecticide-producing or herbicide-resistant engineered crops or cows
injected with recombinant bovine growth hormone.[14] Vandana Shiva, a
physicist turned agronomist, and director of the Research Foundation
for Science, Technology, and Natural Resource Policy in India, has ex-
ploded the myth that GMOs and other biotechnology crops grown in
monocultures produce more food than traditional systems. She writes:

> Yield usually refers to production per unit area of a single crop.
> Planting only one crop in the entire field as a monoculture [the way
> GMOs are grown] will of course increase its yield. Planting multiple
> crops in a mixture [traditional agriculture] will have low yields of in-
> dividual crops, but will have high total output of food. The Mayan
> peasants in the Mexican state of Chiapas are characterized as un-
> productive because they produce only two tonnes of corn per acre.
> However, the overall food output is twenty tonnes per acre. In the
> terraced fields of the high Himalayas, women peasants grow jhang-
> ora (barnyard millet), marsha (amaranth), tur (pigeon pea), urad
> (black gram), gahat (horse gram), soybean (glysine max), bhat (gly-
> sine soya), rayans (rice bean), swanta (cowpea), and kodo (finger mil-
> let) in mixtures and rotations. The total output, even in bad years, is
> six times more than industrially farmed rice monocultures.[15]

Charles Benbrook, former executive director of the Board on Agriculture of the U.S. National Academy of Sciences, said, "Asking too

much of any genetic, chemical or biological technology is a near universal recipe for trouble. Biotechnology does not change this reality. Ironically, the benefits delivered by a transgenic crop variety in a given farming system will often be inversely proportional to the scope of change brought about by it."[16] Miguel Altieri, pioneering agroecologist, entomologist, and professor in the Department of Environmental Science Policy and Management at the University of California, Berkeley, cites U.S. Department of Agriculture, University of Nebraska, and other studies that show that even in monoculture most genetically engineered crops do not yield more food, and many yield less than conventional varieties:

> Based on a cross-sectional examination in 1998 of 365 crop fields in Iowa, GE [genetically engineered] crops provided farmers with no significant difference in economic returns. Yields from GE crops were slightly lower than from conventional varieties.... [In Nebraska], five different Monsanto soybean varieties [were grown] together with their closest conventional relatives, and the highest-yielding traditional varieties in four locations around the state.... On average, researchers found the GE varieties—though more expensive—produced 6 percent less than their non-genetically engineered near relatives, and 11 percent less than the highest-yielding conventional crops. Reports from Argentina show the same non-yield enhancing results with HT soybeans, which universally seems [*sic*] to exhibit yield drag. In fact, the yield penalty that accompanies the cultivation of herbicide-resistant [GE] crops can range up to 10 and 30 percent and is rarely factored into a farmer's calculations in advance.... Yield losses are amplified in crops, such as Bt [GE] corn, where it is mandatory for farmers to leave 20 percent of their land as refuges made up of nontransgenic corn [to prevent the development of pest resistance to Bt insecticide, which occurs if GE corn is grown in extensive monocultures].[17]

The halo of pretense surrounding the alleged benefits of genetic engineering in agriculture is nowhere more evident than in the intense advertising and hype about "golden rice."[18] Vitamin-deficiency-induced blindness is a problem associated with hunger and starvation that threatens more than two million children worldwide. The solution to the problem is a simple one according to the agricultural biotechnologists—engineer more vitamin A into the rice. Recombinant golden rice is the result—it has enhanced levels of beta-carotene, a precursor of vitamin A. Millions of advertising dollars have been spent portraying golden rice as a benevolent triumph of genetic engineering, the answer to the problem of vitamin-deficiency-induced blindness. But like many simplistic approaches to complex problems, this solution doesn't work. Miguel Altieri, explains why:

> People do not present with vitamin A deficiency because rice contains too little vitamin A, or beta carotene, but rather because their diet has been reduced to rice and almost nothing else, and they suffer many other dietary illnesses that cannot be addressed by beta carotene, but that *could* be addressed, together with vitamin A deficiency, by a more varied diet. . . . A magic-bullet solution that places beta carotene into rice—with potential health and ecological hazards—while leaving poverty, poor diets, and extensive monoculture intact is unlikely to make any durable contribution to well-being.[19]

By health hazards, Altieri presumably refers to the fact that vitamin A has a lower toxic threshold than the water-soluble vitamins; and by ecological hazards he refers to the effects of the heavy amounts of pesticides used in rice monoculture. And there is another serious consequence of the introduction of golden rice and other heavily subsidized and government-promoted modern crops. In adopting these new seed varieties, farmers are abandoning their traditional rice seeds, a massive gene pool containing priceless genes for pest and disease resistance, nutritional quality, and high yield. When abandoned, these rice varieties are irrevocably lost. The Indian agricultural scientist H. Sudarshan has shown the magnitude of the problem:

> Over the last half-century, India has probably grown over 30,000 different indigenous varieties or landraces of rice. This situation has, in the last 20 years changed drastically, and it is predicted that in another 20 years, rice diversity will be reduced to 50 varieties,

with the top 10 accounting for three-quarters of the sub-continent's rice acreage.[20]

One cannot help but think that the vast amount of money spent on developing and hyping golden rice would have been much better used in supporting traditional agriculture, which does work—or even in handing out free multivitamin pills to the afflicted children.

One of the most subtle and insidious kinds of pretending that occurs today is practiced by scientists and engineers who ought to know better. This is the use of mathematical models to "predict" future events. Of course, many mathematical models used by scientists and engineers are quite useful, even necessary. This is especially true of models applied to events in the very near future and models that describe fairly simple systems whose characteristics are well understood and can be described in quantitative terms. For example, there are models that predict the crash properties of automobiles and others that predict the resistance of buildings to earthquakes of various magnitudes. Other useful models include those that predict the growth or decline of known populations of animals or plants that have already been studied. If such models are used properly, with attention paid to their defined limits, they work well.

Unfortunately, the success of these basic models has led some scientists and engineers to extend their use in totally inappropriate ways. Two geologists, Orrin Pilkey and Linda Pilkey-Jarvis, in their book *Useless Arithmetic: Why Environmental Scientists Can't Predict the Future*, provide many examples of this kind of unjustifiable modelling. The senior author is one of the world's greatest authorities on beach and shoreline erosion and is himself a qualified modeler, yet their description of shoreline erosion models in current use is scathing. In a lengthy discussion extending through two chapters, they explain how the current models make unjustified assumptions, cannot deal with unexpected major events such as storms, leave out critical variables because they are hard or impossible to measure, and are ill-equipped to cope with beach-to-beach variation. The results are predictions that are little better than chance, and may be worse.

Why are these models in such wide use? The answer: partly because of the pride of the professionals who have staked their careers on them and partly because of political pressure for authoritative answers that serve some political end. The loss of beach shoreline, especially along the east coast of the United States, has created an enormous demand by worried property owners for beach protection and beach replenishment projects, which cost the government many millions of dollars. Yet a beach that has been restored with the addition of thousands of tons of sand may see all that sand disappear in one day in a storm. Hence the demand for models that predict how long the sand will last over the course of twenty-five to fifty years. Never mind that such a model can have no basis in reality; if it can be given as expert evidence in court, that's all that's sometimes needed. Pilkey and Pilkey-Jarvis point out that "long-term monitoring [of the models' predictions] is almost non-existent," as is "meaningful objective comparison of model predictions and the results." Fortunately for the expert modelers, they note, "the public has a very short memory about the projections of beach life spans." In a withering comment that has aroused the ire of some modelers, they state:

> The fallibility or reliability of global change models is widely and openly discussed. Those who invented the beach models, however, express no such concern, brook no criticism from the outside world, enter into no debate about the models, encourage model use under almost any circumstances, and consider skeptics to be unworthy of a response. The coastal engineering profession is a disgrace.

If coastal engineering were the only case of inappropriate model use, it would not be so bad. But this practice is far more widespread. Perhaps the most egregious example given by the authors is the set of models used to justify the long-term storage of high-level nuclear waste at the proposed repository in Yucca Mountain, Nevada. There is urgent pressure from the nuclear power industry and the government to prove, using their models, that this repository is safe; and this has involved the modelers in fantasy rather than science. Pilkey and Pilkey-Jarvis give a number of reasons why the composite model is utterly without validity:

- an impossible predictive time span (hundreds of thousands of years)

- no experience basis for understanding the role of time in chemical reactions of waste or in the degradation of waste containers and the titanium shield
- climate change, a critical factor and a complete unknown
- the magnification of errors created by the interdependence of hundreds of models
- ordering complexity of the natural processes that are involved

What is the alternative to the use of impressive-seeming scientific models that have no validity? The authors suggest "adaptive staging," the process being used in Belgium in designing that country's own nuclear waste repository. In adaptive staging, there is a goal (safe, long-term storage of nuclear waste), but instead of planning the whole process before it starts, based on worthless predictions of events thousands of years in the future, the work proceeds in stepwise phases, with reevaluation at the end of each phase. "[This] would allow the design process to evolve . . . [and] would permit the inclusion of new technology, new ideas as they come forward." But as things stand now, the Yucca Mountain model results are "a fig leaf providing cover for the politicians," a distasteful and dangerous form of pretending.[21]

The last case study, involving pretending on a grand scale, is the so-called antimissile defense shield, dubbed "Star Wars." In 1983, President Ronald Reagan first suggested that incoming nuclear missiles could be rendered "impotent," an idea that grew into the Star Wars missile defense system. If we can send rockets to the moon, a great success, why not use rockets (or lasers) to shoot down other rockets? The idea was believed in by a majority of American citizens, and by 2000 more than $60 billion had been spent on antimissile defense research, with a string of test failures and no working system to show for it. By 2007, after many more billions of dollars spent, there was still no working system, only utopian promises. Indeed, many information scientists and physicists are convinced that such a system is impossible. They think it extremely unlikely that the computer program that will control the shield, one of the largest and most complex programs ever written, could function flawlessly when it was needed. Complex computer programs don't run smoothly the first time they are tried, yet for obvious reasons, this one cannot be tested and debugged except during actual attack conditions. Moreover, in 2003, the study group on missile defense of the American Physical Society concluded that the Star Wars strategy of intercepting ballistic missiles in the early phases of their flight— by rockets or airborne lasers—was essentially impossible for a wide variety of technical reasons. Despite the billions spent on this project to date, the interceptors still have trouble distinguishing between missiles and cheap but sophisticated decoy balloons. Why is a substantial part of the American public willing to accept that a limited but fabulously expensive shield against nuclear missiles is possible, when it manifestly is not and probably never will be? To hide the disappointing failures of the antimissile system, recent tests have been rigged by the Pentagon, using poor-quality decoys that bear little resemblance to real missiles, and by providing the interceptor system with detailed information on "enemy" missile launch time and flight path. Even these rigged tests fail.[22] As Philip E. Coyle III senior adviser to the Center for Defense Information and former U.S. assistant secretary of defense has written:

> Russia has so many ICBMs [intercontinental ballistic missiles] it can overwhelm even the most futuristic missile defenses the United States can imagine, which is why the U.S. Congress voted to shut down the Safeguard Anti-Ballistic Missile system in the 1970s, just one day after it was declared operational. . . . If the MDA [Missile Defense Agency] will not acknowledge that missile defenses are not effective under realistic operational conditions, pretending that U.S. missile defenses actually might work in an all-out war, then they are also pretending that those U.S. missile defenses might work against Russian missiles.[23]

In 2001, the Computer Professionals for Social Responsibility organization awarded its Norbert Wiener Award to Theodore Postol and Nira Schwartz for exposing Star Wars as "expensive science fiction." Postol, an award-winning physicist and nuclear engineer who is professor of science, technology, and national security policy at the Massachusetts Institute of Technology, and who worked in the Reagan administration on antimissile systems, has said that the Pentagon's Missile Defense Agency and its contractors have "misrepresented or distorted" test results and "rigged the follow-on test program." His contention is that the missile defense system can never work, and that Pentagon officials are lying to cover their failures.[24] And Dr. Nira Schwartz, a former senior engineer at the military contractor TRW, lost her job and was investigated by Congress after challenging the company's claims that its weapons could distinguish warheads from decoys. Schwartz, a physicist, engineer, and computer expert with twenty-four U.S. patents to her credit, said: "We've wasted a decade and billions of dollars in a quest for a missile defense shield based on a technology that will never work."[25] Ironically, all of the Star Wars pretending will be irrelevant if, as seems distinctly possible, it is technically feasible and inexpensive for a rogue nation to smuggle nuclear, biological, or chemical weapons into American cities without using missiles at all.

Pretending is a weak thread that is woven into the fabric of contemporary existence, damaging our lives and prospects in many ways. Pretending can squander our wealth, as shown in the examples of expensive projects based on fraudulent long-term models, or the spectacularly expensive Star Wars. It can expose us to dangers from the side effects of our technology, as is happening with GMO crops. Perhaps worst of all, pretending can keep us from pursuing the real solutions to our problems—nowhere is this more evident than in the case of agricultural biotechnology, which is siphoning off scientific brains and money that could be used to further improve traditional farming systems that have already shown their productivity, reliability, and durability.

My case studies of pretending are all drawn from science and technology, because this is what I know best. But the dangers of pretending, of uncritical belief in our own stories, apply equally to the fields of economics, education, philanthropy, law enforcement, and many others. As I wrote in *The Arrogance of Humanism*, "There are limits to the knowledge and power that human beings can muster for any purpose."[26] Ignoring these limits has a seductive appeal, but if we don't wake up soon, many of the opportunities that remain to bring real help to a world in trouble will have vanished, and all our pretending will bring us no comfort.

One theme that occurs repeatedly in the chapters that follow is that we have to stop pretending and start being honest with ourselves. This morning, a beautiful day in early June, I watched a team of men from a "lawn service" company working on a neighbor's postage-stamp-sized lawn. First they mowed it with machines big enough to cut the Amazon rainforest, then they sprayed it with toxic chemicals to kill the dangerous dandelions and violets, and finally—the *pièce de résistance*—they cleaned up with gasoline-powered leaf blowers that directed violent blasts of wind onto the customer's sidewalk, steps, and driveway. The noise echoed through the neighborhood as huge clouds of dust and particles rose up into the air, some of it remaining airborne while the rest settled back in the open windows of the house and on all the neighbors' properties nearby. A few long-handled brushes, I thought, would have been cheaper, more effective, and far less damaging to health, peace, and property.[27] At first I fumed, then realized that it was silly to get angry at what I was watching. So I focused on the absurdity of the scene, remembering Mr. Bennet's famous remark to his daughter Elizabeth in *Pride and Prejudice*: "For what do we live, but to make sport for our neighbors, and laugh at them in our turn?"

But of course there has to be more to life than that. In the tales in this book, you will find the follies, both amusing and grim, that now mark so much of our technology, our economic practices, and our relationships with nature, with our children, and with the other people around us. More important, you will also find guideposts pointing the way to the grand possibilities inherent in a world in which permanence, care of others, care of the land, and good work are again qualities that are prized above all others.

---

## BRAINSTORMING HAS ITS LIMITS

Thinking is the most overrated human activity.

Wendell Berry[1]

A number of years ago, I participated in a brainstorming session in Washington, D.C. Although not my first panel meeting, it was, nevertheless, the first that I served on that was officially called a brainstorming session. I was filled with nervous anticipation as my train pulled into

Union Station the night before. The next morning my anxiety increased as I sat munching Danish pastries and drinking decaf from a disposable paper cup in the posh conference room of the environmental organization that was cosponsoring the meeting with a federal environmental agency. What was brainstorming? Would I be able to do it?

I had looked up "brainstorming" in the *Oxford Modern English Dictionary* back home in New Jersey. The first definition given was "a violent or excited outburst often as a result of a sudden mental disturbance." This didn't seem right. Nor did the second definition: "*colloq.* mental confusion." Not until I got to the fourth definition was there a glimmer of hope; it read, "A concerted intellectual treatment of a problem by discussing spontaneous ideas about it." But I still felt confused. What was a "spontaneous idea"—or, to put it another way, what was a nonspontaneous idea? How did a discussion of spontaneous ideas differ from an ordinary serious discussion? Hesitantly, I had concluded that a brainstorming session must produce more exciting and definitive results than a mere discussion.

At first I was alone in the conference room, free to eat more than my share of Danish. Then a woman arrived. By her air of confidence and authority I knew instinctively that she was the chair of the session. We spoke, and I casually turned the conversation to the topic of brainstorming, maneuvering to get some kind of clue about what was expected of me. But she turned out to be the audio technician, more interested in the technicalities of the recording system. No hope. Soon the room filled with august biologists and social scientists; I sank inconspicuously into my high-backed chair, trying to breathe softly and blend into the upholstery pattern.

The session began. Our charge, as the expert panel on biodiversity, was to determine the "megatrends" that will affect the global environment in the next thirty years, and to predict which "events," "forces," "uncertainties," and "surprises" would have significant environmental impacts during that time. Right from the beginning I fell behind while musing about how one could predict surprises, and when I snapped out of it, I became aware that I was lost. The conversation didn't make sense to me—I had missed too much. Each panel member seemed to be talking about his or her specialty or personal experiences; I couldn't make out the subtle connections that were no doubt weaving the individual

strands of conversation into a brilliant, unified prediction of our biospheric destiny.

Time passed. I ate more Danish and found that I was assigned to one
of the three or four working groups into which the expert panel had
been divided. Now we were writing questions, trends, forces, and surprises on a huge pad of paper held on an easel. The colored markers we
were using had a toxic smell: I blamed my mental confusion during this
part of the brainstorming session on the fumes from those markers.

The final two hours of the meeting were devoted to reconciling the
findings of the various working groups. Because of the markers, I didn't
understand my own group's conclusions; soon I discovered that I couldn't
follow those of the other groups either. They appeared to be statements of
the obvious, and it was then I realized that the effects of the fumes had
not quite worn off. We were into the moment of grand synthesis, the final
spurt, and I was still floundering near the starting line. The best I could do
was nod sagely and let the flashes of wit and insight pass me by. When the
brainstorming session ended, I left quietly, assuming that I had done a
pretty good job of concealing my inadequacy. I was wrong.

An envelope containing the draft summary report of the meeting
arrived in my office a few weeks later. I opened it eagerly. The report
came to the point quickly. On the second page of text I saw four "strate-
gic issues" printed in bold type—the first read: "The future of biodiversity will be determined by social, cultural, and historical forces." With a
sinking feeling, I quickly scanned the other three bold entries. They were
just like the first. The fourth stated, "A 'patchwork quilt' of constantly
evolving institutions is required that can solve problems at various
scales." This was followed by many pages of text and tables, including a
whole page of possible "negative" and "positive" "surprises," the first of
which was "meteorites." Where were the exciting and definitive results
that I had expected? The report of the brainstorming session was of no
practical use to anyone.

I felt sure the results were my fault. I had failed; I had not pulled my
weight. I had ruined the brainstorming session. Evidently, for brainstorming to work everyone must be having spontaneous ideas, and my
ideas had hardly been spontaneous at all.

Of course, once the word about me got out it would be unlikely that
I would ever be invited to another brainstorming session, but there was

always a chance, and the next time I must be prepared. What was I doing wrong? I used to believe it was best to tackle environmental and other problems that I had a personal stake in, to confront them directly one at a time, and to concentrate mostly on the present where I was on firm ground, even though real progress was often slow. This way, I thought, there were time and opportunity for personal knowledge and personal emotions to inform one another, minimizing the errors brought on by pure thought or pure feeling. Here was my mistake; I must have had a bad attitude. I hadn't realized the power of expert planning and forecasting, the importance of thinking globally, and the incredible possibility of figuring out the earth's future after one or two days of exchanging spontaneous ideas around a conference table— given the right mix of experts and facilitators.

It was too late to fix the session that was past; however, I could improve my brainstorming skills by replaying my role at the expert workshop in the privacy of my office, trying to get it right. For nine hours, without pausing for food or drink, I practiced having spontaneous ideas about the megatrends that will affect the future of biodiversity, technology, and society. Hunger and thirst sharpened my faculties (I thought ruefully of the Danish and decaf), and after several hours, ideas began to pop into my head almost too fast to write down.

I thought about the device almost everyone will be wearing in the coming electronic age that will alert them when they are about to make the transition between virtual and ordinary reality: a device with vinyl bladders, filled with dried, simulated peas, that flap the wearer gently on the ears—left ear when going into virtual reality, right ear when going out. I thought about the dawning era of global free trade and the single remaining multinational corporation, Consolidated Z, which will control all products, prices, wages, and local zoning variances. . . . The ideas were still not quite spontaneous, and there was an unsettling residue of the specific. But I was improving steadily.

Around six in the evening, a hunger spasm made me turn my head, and my eyes were distracted by a copy of an essay by the British philosopher, Mary Midgley, titled "Why Smartness Is Not Enough," lying on a corner of my desk. It was open to page 48, where I read the words, "All around us, we can see people trying to solve by logical argument, or by the acquiring of information, problems that can only be dealt with by a change of heart."[2] For a moment I felt discouraged—had I been wasting my time in a fruitless activity? Then I stretched, took some deep, aerobic breaths, and was myself again. Midgley must have been referring to other people, not to brainstorming environmentalists. Time for one more spontaneous idea.

I thought, finally, about the last remaining wood turtle, in the spring of the year 2019, an old barren female in Potter County, Pennsylvania, tapping the moist earth with a brick-red foreleg, her wise and patient brown eyes searching for the first sign of a responding worm, while on the ridge top the yellow bulldozers gather and miles away, in Washington, the experts are still brainstorming, brainstorming.

---

## NOTHING SIMPLE

A saying attributed to the eighteenth-century Ukrainian philosopher Gregory Skovoroda has long stayed in my mind. Skovoroda wrote that "we must be grateful to God that He created the world in such a way that everything simple is true and everything complicated is untrue." I have always suspected this statement to be profound, and I would like to believe it. I often do believe it. As I write these words, I am pretty sure that I believe it. But I can't say that I am 100 percent free of doubt. Gregory Skovoroda came of simple peasant stock and lived in the days before multinational corporations, environmental management, lengthy tax returns, and global free trade. I know that great wisdom is supposed to be timeless, unaffected by changing circumstances, yet I wonder if exceptions can be made for times like ours. The trouble is that so many of the things I used to think were simple have turned out to be complicated. At times, usually around six in the morning when I am still in bed and life seems grim, I even wonder whether anything is simple anymore.

Take the case of overpopulation. Any ecologist worthy of the name knows two things about overpopulation: first, it is bad; and second, the United States, because of its excessive per capita consumption, is already overpopulated. In the United States and all over the world, ecological, social, political, cultural, and economic systems have all been strained past the breaking point by the burden of too many people consuming too much. The connection among population, technology, and ecological degradation has been thoroughly demonstrated in countless ways, perhaps most formally by the Millennium Ecosystem Assessment, a project enlisting the help of more than 1,300 ecologists and other researchers from 95 countries.[1] And the human population, at more than 6.5 billion at this writing, continues to grow. Although fertility rates are declining throughout the developed world and are predicted to begin to decline in the developing world in the years ahead, the earth's population is still predicted to grow to approximately 9.1 billion by the year 2050. True, a host of disasters—food supply, disease, ecosystem instabilities including climate change, social upheavals, war—may cut short population growth, even cause a decline. But at the moment it looks like the earth will have wall-to-wall people.[2]

Most citizens do not yet perceive the magnitude of this danger. The root causes of our worst problems rarely appear in plain view. As the biologist and social critic Garrett Hardin once wrote, "Nobody ever dies of overpopulation"; the death certificate—if there is a death certificate—lists something more tangible such as cancer, tuberculosis, or, most commonly, pneumonia.[3] It is difficult to acknowledge distant ultimate causes such as overpopulation when there is something easier and closer at hand to blame. Nevertheless, increasing numbers of people, even including some economists, are beginning to recognize the simple idea that the world and its natural wealth are finite and that population growth must stop.[4]

But are things really so simple? One complication is that the perception and judgment of overpopulation seem to be culture-dependent. National comparisons do not work when population is concerned. Take China and the United States, for example. In the United States, with more than 300 million people, many are feeling the stress of crowding—in the national parks and in the expanding urban conglomerates, for instance. So, not surprisingly, are the 1.3 billion people of China.[5] But what was China like in the early decades of the last century when it had one-fifth of its present population, 300 million? With less habitable land in China than in the United States, was Chinese overpopulation in the 1920s and 1930s similar to that in the United States in 2007? I doubt it.

The Chinese, at that time, ate relatively little meat and had, compared with Americans, few large farm animals and little of the massive waste of calories and protein that occurs when grain is fed to cattle, so their efficiency of food production was much higher than ours. Nor did the Chinese ever use as much energy per capita in keeping themselves clothed, housed, and amused. And China, as befits a country that has been densely populated for a very long time, over many centuries has evolved customs and codes of social behavior that make it possible for large numbers of people to rub elbows without society unraveling in chaos, arson, and bloodshed.[6]

Prominent among the cultural traits that have enabled large populations to live together in China has been a long-standing sense of self-discipline combined with an awareness of the narrowness of the social pyramid as it tapers toward its apex. This applied to the privileged as well as the lower classes. Until the early years of the twentieth century, a Chinese boy of a cultivated family was taught his first twenty-five written characters shortly after his third birthday. They spelled out an encouraging but cautionary verse:

> Let us present our work to father.
> Confucius himself
> taught three thousand.
> Seventy were capable gentlemen.
> You, young scholars,
> eight or nine!
> Work well to attain virtue,
> and you will understand propriety.[7]

By the time he was fifteen and ready to take the imperial examinations for a government position, the boy had memorized texts containing 431,286 characters. (Meanwhile, his sister was learning a different and even more demanding kind of discipline and submissiveness to male authority.) These imperial exams, which were instituted in the sixth century, continued to be given until 1905. Echoes of the system persist in contemporary, densely populated Japan and in other parts of crowded Asia. In the United States, by comparison, fewer social adaptations to crowding seem to exist. As we push on toward a population of

310 million people, we still retain a fixed idea of personal space so grandiose that it was wasteful and destructive even in frontier days, when there were fewer than a third as many Americans as there are now. This is the kind of intolerance of crowding that actually led citizens of New Jersey during the "energy crisis" of 1973 to murder each other for a better place in the gas station lines, and which has reappeared in other forms since then. China today faces some of the most overwhelming environmental problems in the world, especially air pollution and the dwindling supply of unpolluted water for drinking and agriculture. But should America, with its much smaller population, face the same challenges, I doubt that it would cope as well because of the lack of socially engrained restraints.

If, as seems likely, the definition of overpopulation is complicated by national relativism, with each country or culture experiencing a different threshold, how can we say that this is a simple and therefore true idea? Would Gregory Skovoroda, having followed the argument up to now, turn his back and walk away in disgust? Or could we persuade him that the basic idea of overpopulation is still a simple one? "Gregory," we would say to him, if American, or "Mr. Skovoroda," if Chinese, "wait a minute. Don't be in such a rush to leave. Overpopulation is not a complex notion even if there are a few qualifications. After all, what could be simpler than the statement that we need one-child families for the world to survive? One child, Gregory [or Mr. Skovoroda]—that's what the world needs. You can't get any simpler than that." Or can you? Leaving the Ukrainian philosopher to ponder our words by himself, let's step out of earshot and look more closely at the simple idea of the one-child family.

When I was growing up, I had a younger sister, but a four-year difference in age, plus sibling rivalry, put her beyond the pale for companionship as far as I was concerned. Fortunately, for I was socially backward and had few friends, I had cousins in the neighborhood whom I enjoyed playing with. My favorite was Laney, who was almost my age and who in addition to being lively and interesting lived around the block from me. Our backyards touched at one corner, making visits convenient; although to get from one yard to the other in the easiest way we had to cross through the flower beds belonging to an elderly couple who yelled very loudly at us when we trampled their plants.

Laney and I taught each other to play table tennis using a kind of elongated card table with a makeshift net and, in lieu of real paddles, the hard backs of books that we had salvaged from the trash. I have never had as much fun using proper equipment. Laney is now a professor of statistics at an East Coast university. I don't know whether she still plays table tennis. I also remember the parties, mostly at cousin Bobby's house, where Bobby, Jimmy, Laney, and occasionally Billy, who was older, all vainly tried to help me conquer my fear of girls.

My childhood was enriched by cousins because my father happened to have four married siblings living nearby. My cousins on my mother's side lived in faraway places such as Minneapolis and Miami—I rarely saw them. But my parents and all of their surviving siblings, a total of nine brothers and sisters, were extremely close to one another regardless of the physical distance that separated them, sharing happiness and supporting each other in every way during the inevitable times of trouble. This sense of family was the dominant influence during my early years.

In traditional China, family was of even greater importance. Not only were ancestors worshipped and elders respected, but relatives of every age and degree were recognized and significant in each person's life. As Charles O. Hucker wrote in his history, *China's Imperial Past,* "The individual was never alone in his confrontations with the world at large; he was representative of, responsible to, and supported by his family."[8] The Chinese language itself has words for various kinds of family relationships that take whole phrases to describe in English. Even in modern China, this dominant fact of the family as the center of social existence has persisted in spite of the one-child policy begun under Mao.

And what of the one-child family? In a nation of one-child families, what words drop out of daily language? "Brother" and "sister" to begin with—there are no brothers and sisters in one-child families. Because there are no brothers and sisters, there can be no "aunts" and "uncles," and therefore no "cousins," neither first nor second nor once or twice removed. No "sisters-in-law," no "brothers-in-law." Blood may be thicker than water, but in a one-child world, water will have to do.

Of course many people grow up happy and well without siblings, cousins, aunts, or uncles. And some extended families are stifling and terrible. It is perfectly possible to have good, close relationships with nonrelatives during childhood and afterward; nearly everyone does. But an extended family can offer a richness—and sometimes a stability—that is hard to duplicate in modern societies, whether affluent or impoverished. In affluent societies such as ours, most social forces seem to move the individual toward an ever-greater isolation that is masked but not compensated by the riotous consumption of material goods and the growth of electronic communication. Families can be the best antidote to isolation. In poor societies, an integrated family is often the only barrier between an individual and political oppression or death from starvation.

Thus the passing of the large family, necessary as this may be, is not the simple demographic blessing that the ecologists and population planners would have us believe. What will become of us without it? Social evolution, normally slow, may under the pressure of necessity replace the traditional family with new kinds of social structures based on an average of one child per mother. For example, single adults, childless couples, and parents with various numbers of children might be part of a formal "family" that offers support to all. Indeed, such families already exist, and if we are lucky, they—or some other modified social structures—will turn out to be sustainable in a changing world. But none of this is yet clear.

In the meantime, the human population is growing and growing. The earth groans under it. There is no place to hide. Less and less is simple. If Gregory Skovoroda is right, everything that we believe about our society and our future becomes untrue because it is so complicated. Or perhaps I have misapplied or misunderstood his words . . . perhaps he wasn't referring to things like overpopulation and other modern problems . . . perhaps some meaning was lost in translation from the Ukrainian. . . .

---

## THE COMFORTS OF FANTASY

Some people have a drink when they get home from work—a martini, a beer. Maybe two or three. Life is especially stressful in the twenty-first century. The same indecent forces that are destroying nature are disrupting our working lives as well. Who will own the company tomorrow? Will there be a "reduction in force" or some other euphemism for the ax? When is the next reorganization coming? Are my ten or twenty years of faithful service an asset or, more likely, a sign of obsolescence and suspicious loyalty to bosses and coworkers now out of favor or working for other companies? How am I to answer the e-mail message that arrived at 4:00 P.M., the one that seemed to contradict the one that arrived at 11:00? When will I find time to fill out the questionnaire from the Resource Management Office, and does it take precedence over the Goals Enhancement Strategy questionnaire that came from the Administrative Services Division? Are my computer files compatible with the new software system, and, if so, why did the box on the screen say, "This paragraph is unreadable. Do you wish to substitute a standard paragraph?" Just what is our real work, anyway?

Alcohol can take the edge off stress but it is not everyone's consolation. Some choose television or the Internet; mine is to pick up one of the scores of detective novels that I keep close to hand and plunge in. Then I can forget for a little while the vice presidents, deans, and other academic, corporate-style bosses who do their best to make life in the modern university an unproductive misery. In this way I can put out of mind, temporarily, the pleas of students who don't understand why there are no courses to take and the ravings of colleagues who can't figure out how to cope with the contradictory, impossible demands placed on them.

Why should detective stories be, for so many, such a good and entertaining way of escaping from reality? That they are is clear; billions of copies have been sold. What's more, the fictional detectives endure decade after decade, outliving their creators, even their creators' children. In the real world as time passes, gas lamps give way to carbon arc lights, which are replaced by tungsten bulbs, which are supplanted by mercury vapor lights, halogen bulbs, LEDs, and a host of fluorescents; yet Sherlock Holmes still glides through the lamplit streets of Victorian London as the wisps of yellow, pea-soup fog gather and drift in countless bedrooms on six continents.

Modern imitations of Conan Doyle's Holmes and Rex Stout's Nero Wolfe—some actually bearing the names of these immortals—claim valuable shelf space in today's bookstores, but copies of the original stories outnumber them. The character of the great Chinese detective, Judge Dee, embellished and reinvented by many centuries of Chinese mystery-storytellers, and in the twentieth century brilliantly resurrected and re-created by Robert van Gulik, has survived since Dee's actual death during the Tang Dynasty, at the end of the seventh century. Hercule Poirot still twirls his waxed mustache and sips sickeningly sweet sirops, outselling the competition in the days of purple hair and margaritas. And the Hardy boys and Nancy Drew, though they are now no strangers to jet planes, remain most compelling and most popular in books written when gearless bicycles and ancient Fords were the only transportation available to these adolescent sleuths.

We may feel that our stress is greater than that of past times; indeed, it is only during the last sixty-plus years that we have developed the most destructive technology and the most devastating overpopulation in human history. But the detective story, with many of the same detectives who comforted our grandparents during their troubles, is still capable of comforting us. This calls for an explanation. I have one.

What the detectives of fiction offer to us is confirmation by example of an idea that is becoming ever more in conflict with the reality of our lives and ever more precious: the idea that there is an order and sense to the happenings around us, and that this order can be discovered if we know how. We have the possibility of understanding and controlling events. Television and the Internet give us similar illusions.

The detectives' secret, at least for the best of them, is simple, elegant, and—once you know it—elementary. To understand the inner workings of our complex, seemingly chaotic surroundings, the great detectives move outside of society, retaining only a narrow, carefully defined and managed connection with the world of unfathomable events and confused people.

Sherlock Holmes established the pattern. When we and Dr. Watson first meet him, in *A Study in Scarlet*, he appears at once to be an odd sort: "A little queer in his ideas" and "desultory and eccentric" in his studies, he has nevertheless "amassed a lot of out-of-the-way knowledge that would astonish his professors." As Watson's roommate, Holmes proves fascinating and occasionally congenial, as long as the infinitely obliging Watson respects his privacy and the arrangements he has made to keep the world at arm's length. Watson has amazing tolerance for Holmes's moods, both manic and depressive, for his use of cocaine, for his secrecy, for his impatient arrogance, and for his apparent lack of ordinary human emotions. This last point needs some qualification. In a handful of instances, perhaps most obviously in "The Adventure of the Three Garridebs," when Watson receives a flesh wound from a bullet, Holmes does fleetingly betray an affection for Watson, although much of the time he behaves in a way that today might earn him a diagnosis of borderline Asperger's syndrome.

The lure for Watson, apart from that curiously strong kind of friendship that can occasionally spring up between opposites, is the same lure that attracts us. Holmes finds sense and pattern where we see only a meaningless jumble. From a battered hat found in the street, Holmes deduces that the unknown owner is an intellectual gone to seed during the past three years, middle-aged, sedentary, out of training, whose wife has ceased to love him, etc., etc. With the passage of more than a century since "The Adventure of the Blue Carbuncle," many of Holmes's deductions seem improbable or preposterous to our sophisticated eyes. But, like Watson, we are still happy and flattered to spend time with him.

With Nero Wolfe and his assistant, Archie Goodwin, Rex Stout took the Holmes concept and, in the four decades following 1934, revised it for the age of atomic bombs, television, administrators, and consumers. Stout's achievement, a monumental one, was to show that it might be possible for a person to be both successful and free in our society; he demonstrated—if only in fiction—the power that can come from skillful and uncompromising avoidance of the entanglements and commitments of commercial culture. To do this, Stout moved Wolfe farther from society than any detective before him.

The big difference between Nero Wolfe and most of us is that Wolfe sets his own daily schedule and orders his own life. Self-employed, naturally he determines his hours, but I don't mean that he works half-days or takes Tuesdays and Thursdays off. His day begins around eight, when his cook, Fritz Brenner, brings breakfast to his room. From nine to eleven in the morning and from four to six in the afternoon, Wolfe is in the greenhouses on the roof of his Manhattan brownstone, supervising his gardener, Theodore Horstmann, in the care and breeding of a vast collection of choice orchids raised for his amusement and pleasure. If clients try to see Wolfe while he is in the plant rooms—no matter how urgent their business or how influential they are—Archie Goodwin sends them away or makes them wait. Likewise for district attorneys and police inspectors. Murder and mayhem may swirl around his doorstep, but at the times he has appointed for himself, Wolfe remains serenely ensconced among his orchids.

Nor is any business discussed at meals, which are long and lavish (the Nero Wolfe stories have spawned at least one cookbook). Wolfe refuses to leave his house on business and seldom leaves for other reasons; he prefers the society of his household and a few close friends, and, moreover, distrusts and loathes cars and planes. He rarely shakes hands, never makes small talk, and takes on cases only when it suits him and he needs the money. He does not rely on gadgets and technology to solve crimes; his knowledge of what is going on in the world comes primarily from newspapers, not television.

Wolfe is massive (one-seventh of a ton, we are told), formidable, and even less given to expressing emotion than Holmes. There is no way to relate to him except at his discretion, on his terms, and then across a barrier. The only people he respects are those who, as he does, define themselves without regard to outside influences. Nero Wolfe dictates his own needs—the ultimate antimodern man. In Wolfe, Rex Stout has created that rarest of types in our time, the true conservative, one who insists that there are unchanging standards by which all things, from food to progress, must be judged.

The fatness of Wolfe—also a rejection of our society—is as principled and deliberate as the other parts of his existence. In his house, the preparation and eating of food is a cultural statement, religious in intensity and serving some of the same separating and identifying purposes that food serves in traditional Jewish, Hindu, and Islamic life. Once, in the book *In the Best Families*, Wolfe is forced to flee his house and change his identity. He becomes part of the culture he spurns, subsists on the typical American food he detests, and loses a huge amount of weight. His acquired thinness is more than a disguise—it is a way of impressing the reader with the magnitude of his sacrifice and the connectedness of the different parts of a principled existence. Once the threat is removed and he can go home, Wolfe's first act is to call the farmer who supplies him with poultry and instruct him to resume feeding select chickens on blueberries. Only then do we know that things are back to normal on West Thirty-fifth Street.

It is Archie Goodwin, Wolfe's legman, who connects Wolfe with the world around him (and with us), making the detecting plausible and the books so much fun to read. Archie revels in the culture and amenities that Wolfe rejects—the fast cars, the fashionable, expensive suits, and the baseball games. To say nothing of the dancing at the Flamingo Club. (The outspoken misogyny of Holmes and Wolfe is another way of setting them apart from the world and of making things simpler and therefore more fathomable. It is consistent with their characters, but it is not a necessary ingredient in detective fiction.) Archie enables Wolfe to manipulate and relate to a society that Wolfe chooses not to enter. Wolfe needs Archie, but what allows Wolfe to tolerate him is Archie's integrity of character. Like Wolfe, Archie is self-defined; he sets and keeps his own standards. The two men understand what it takes to live uncorrupted in the modern age, so they understand and appreciate each other.

Both Nero Wolfe and Sherlock Holmes are answerable first to themselves. If their principles coincide with the law, as they usually do, they uphold it. If not, they get around it. In "The Adventure of Charles Augustus Milverton," Holmes and Watson break into a private home, where they deliberately destroy evidence that would identify a murderer whom they believe should not be punished. In *The Next Witness*, Wolfe evades a subpoena and an arrest warrant to find a criminal his own way. There are many other instances from the accounts of these detectives and others. A notable example is *Mr. Pottermack's Oversight*, by R. Austin Freeman, in which the scientific detective, Dr. Thorndyke, solves a murder that has baffled the police; he then decides not to tell them who did it. In all such cases, the detectives share the perception that fidelity to people, places, and principles can be more important than the formal requirements of law, and that the law is not a license to suspend personal judgment and responsibility.

Both Conan Doyle and Rex Stout, like the celebrated detectives they invented, were originals, people of exceptional integrity and talent who had proven to their own satisfaction that it was possible, with skill and care, to change the world around them. Doyle was, besides a writer of fiction, an ophthalmologist, successful inventor, dramatist, war correspondent, whaler, athlete (he is said to have introduced skiing to Switzerland), historian, and spiritualist. Stout, a child prodigy in arithmetic, was a naval officer (on President Teddy Roosevelt's yacht), an itinerant bookkeeper, sightseeing guide, stable hand, hotel manager, inventor of a widely used school banking system, radio commentator, prominent opponent of Nazism and nuclear weapons, cook, and (like his sister, Ruth Stout) a brilliant pragmatic gardener. Nobody could describe them as narrow or unworldly. They each chose to distance their detectives from the worlds of their time, but these were decisions based on wide experience.

In Sherlock Holmes and, especially, Nero Wolfe, I have picked extremes to make a point. Most other fictional detectives move more comfortably through the world about them. But nearly all have some distinguishing characteristic that lets them step away from and make sense of their environments. Here are a few examples of the best: Sergeant Jim Chee (Tony Hillerman), a religiously committed Navajo, is trained in anthropology; Mr. Campion (Margery Allingham), Lord Peter Wimsey (Dorothy Sayers), and Lovejoy (Jonathan Gash) live on the fringes of their respective societies; Jane Austen (Stephanie Barron), a detective based on the early nineteenth-century novelist, is a woman in a man's world; Inspector Ghote (H. R. F. Keating) and Superintendent Otani (James Melville) have extraordinary standards of personal honesty which set them apart; Inspector van der Valk (Nicolas Freeling), unimpeachably Dutch, has a perceptive and embarrassingly unconventional French wife; Brother Cadfael (Ellis Peters) is a Benedictine monk with vows of renunciation to observe. Each of them discovers and restores the underlying pattern in a world that makes sense in spite of its ambiguities, in spite of how chaotic and frustrating it seems to us ordinary folks.

But the world, especially our world of the twentieth and twenty-first centuries, does not make elementary sense, and makes less and less with every passing year. The idea that it can be figured out and manipulated by exceptionally talented individuals is a comforting lie. There are a few modern fictional detectives who know this, who admit and explore the unpredictability, complexity, and uncontrollability of our world, and who in the end must fail. These detectives, like Guido Brunetti (Donna Leon) and Adam Dalgliesh (P. D. James), suspect that, in the words of Dalgliesh, there may be no "great underlying pattern." Their world is our world—a world in which nothing is as it seems and defeat masquerades as victory. The appeal of these detectives is in their ability to see things as they really are, to forego any pretense of control and omniscience, yet to remain human and accessible and to survive. We may not be able to change the world, they seem to tell us, but understanding our place in it is no small feat. The considerable enjoyment these latter detective stories provide us is that of the good novel, not the comforting reassurance of the classical mystery. The difference is real, as Conan Doyle and Freeling discovered when, having killed off their detectives, they were forced by their readers to bring them back again.[1]

Meanwhile, it is pleasing after a day at the job to step silently and invisibly into Nero Wolfe's office, stand next to the great globe, and dream that there really are people like Wolfe who can figure out what is going on. The insatiable demands of global management in our increasingly self-constructed world cry out for a Wolfe, a Holmes, a Poirot—people who can discern the patterns and gather up the myriad, tangled strings that make the marionettes dance. But if we are honest and look around us at the cascading problems caused by our prior solutions to earlier problems, we can see, like Brunetti and Dalgliesh, that no human, and certainly no machine, is up to the task of fixing them single-handedly. We never were, of course, but before the age of science most were willing to admit our ultimate shortcomings.

Could there be even a modern *fictional* equivalent of Wolfe in this disintegrating culture? Yes, I think so, but with changes. What will the next Nero Wolfe be like? For one thing, he will surely disdain e-mail and the Internet. For another, he will probably be a she. I'm waiting eagerly. The twenty-first century takes a lot of escaping.

# TWO

---

KEEPING TRACK OF OUR LOSSES

Civilizations are fragile, impermanent things.

Joseph Tainter, *The Collapse of Complex Societies*[1]

# REJECTING GIFTS

> To forget our mistakes is bad. But to forget our successes
> may be worse.
>
> Neil Postman, *Building a Bridge to the Eighteenth Century*[1]

A long-remembered scene repeated many times: I am about sixteen; it is 5:45 P.M. and we are at the dinner table, with my mother at one end, my father at the other, and my twelve-year-old sister, Judy, sitting across from me. As always, the meal is formal—the oval mahogany table is covered with a clean tablecloth, and there are three courses: appetizer, main course, dessert. And as always we are eating very quickly to keep up with my father, who must be back at the office at six to see more patients. The phone rings; I can hear our maid, Josephine, going to answer it. A pause. Then Josephine comes into the dining room. "It's Dr. Williams," she tells my father.

My mother is annoyed. "Why does he always call at dinnertime?" she asks, but my father is already on the way to the phone. Dr. Williams and my father, both practicing physicians, occasionally meet at Passaic General Hospital, but Dr. Williams, who has heart problems, is also my father's patient, and that is why he is calling. He knows that my father puts his patients ahead of his dinner. Soon, my father returns to snatch a hasty bite of dessert and tells my mother, "I'll be going to Rutherford after office hours tonight. I should be back around ten." My mother says nothing.

Another scene, equally well remembered: My father and I are in a car, and he is driving; I am not yet old enough for a license. He speaks to me in that tentative, half-apprehensive tone of voice that fathers use toward adolescent sons when they are about to propose something new. "You know that I have a patient, Dr. Williams, over in Rutherford? Well, he's beginning to go blind and needs somebody to read to him once a week for an hour or two. I told him I would ask you—he would pay you for your time."

From my hesitation, my father knows that I am not liking the idea. "You would enjoy reading to Dr. Williams," he continues. "He's a very interesting man. He writes poetry." But it's no use. I make my excuses: too busy, too much schoolwork, too inconvenient. The fact is, I just don't want to read to an old friend and patient of my father. A few days later the request is repeated, then a week after that. By this time it has become a matter of principle, and I am adamant. Eventually my father stops asking.

A third scene, just a few days before this writing: Dr. Williams is long dead; my father is dead, too. I am sitting in my living room holding a familiar book from my father's collection. A slender volume, elegantly printed on creamy, rough-cut paper yellowing slightly at the edges, it is the fourth book of a set, which together make up one of the great epic poems of twentieth-century English literature. Opening it carefully to avoid tearing the fragile dust cover, I read that this is the first edition of Volume IV of *Paterson*, of which one thousand copies were printed for the publisher, New Directions, by Van Vechten Press, Inc., Metuchen, New Jersey.[2] And I see once more the inscription by the author, written in blue ink: "Irving Ehrenfeld, gratefully yours, William Carlos Williams, 9/19/51."

Despite the painful regrets I have, I don't blame my sixteen-year-old self for rejecting the marvelous gift offered to me by my father and his celebrated patient. Now, with two recently adolescent sons of my own and two daughters who were adolescent not so very long ago, I understand that becoming independent sometimes involves the rejection of gifts. It is a necessary cost of human development, a tragic cost, because so many of the gifts can never be offered again. But it is a cost that must be borne—there is no alternative.

The theme is one that I think about often because we live in a time when the unnecessary rejection of gifts is commonplace, a true way of life. Our cultural landscape is littered with fragments of gifts rejected unnecessarily, and it is the adults, not the adolescents, who are doing the rejecting. The techno-economic system that dominates the industrial world has little patience for the bearers of evolved wisdom, for anyone encumbered by connections with the past and concerns for the future.

This was brought home to me one day when I met Dr. Oved Shifriss on the grounds of Rutgers's Horticulture Farm No. 3; I hadn't seen him in about a year. Retired from the faculty for more than a decade, he was then in his eighties. The shapeless, faded blue canvas hat was the same as always, but he was now wearing hearing aids in both ears. The senior farm worker who was with him, much younger and obviously very busy, waited patiently and respectfully while we spoke—or rather while Dr. Shifriss spoke and I listened. Oved Shifriss, son of Israeli pioneers, once the chief of research for Burpee Seeds, and probably the greatest vegetable breeder Rutgers has ever had, knew as much about the genetics of squashes, cucumbers, melons, and pumpkins as any person on earth. His creation of Burpee's Big Boy tomato, in 1949, helped ignite the American boom in home gardening. When I first came to the university, I was surprised to notice that even his long-time colleague, himself an accomplished scientist, always addressed him as "Professor Shifriss." I soon understood why.

Shortly before his retirement, Dr. Shifriss told me the reason for his decision to leave. He had recently won a major award in the All-America vegetable trials for his Jersey Golden Acorn squash, but he wasn't satisfied. "This is only the beginning," he said. "If they gave me a team to help me—a few geneticists, one or two plant pathologists, an entomologist, a biochemist—I could produce a squash that would feed the world. It would be about the size of a sweet potato, very productive, insect-resistant, maybe with silvery leaves, highly nutritious, rich in Vitamin A, protein, and calories, and very delicious. I can do it, if the administration would only give me some resources and leave me alone.

"But they don't seem to be able to grasp the idea. You know how they criticized me when I was working with the little ornamental gourds. 'Those gourds are bitter and inedible,' they said. 'Why are you wasting the Experiment Station's money on them?' But I got the B-gene complex that enabled me to breed the Jersey Golden Acorn partly from the same gourds. It tripled the amount of carotene. And look at the trouble they give me because I am doing some studies with castor beans—the plant that Jonah sat under in the biblical story. 'They're poisonous,' they keep telling me, 'what good are they? You're supposed to be a vegetable breeder.' But castor beans have an extremely unusual sexual system; it's going to be important one day."

By the time he retired, not long after that conversation, Dr. Shifriss was spending a sizable part of his own salary to replace the research support that the Experiment Station had withdrawn. The super squash that he envisioned and could have created was never developed; he died in 2004, and there is probably nobody left with his combination of knowledge, experience, and genius to do the job. For a while, genetic engineers from Europe showed some interest in his castor bean studies; I don't know what came of that. In fact, castor beans are why I ran into him at the farm: he was still allowed to keep a couple of his special plants in one of the greenhouses there. We talked a bit more—he was explaining some of the reasons for the high failure rate of genetic engineering in agriculture, and his ideas about the unexpected variability he had seen within cloned plants that are supposed to be identical—then, tiring, he said good-bye and walked slowly off toward the greenhouse.

As I watched him go, I thought of the terrible waste of his talents and wisdom. Our demand for immediate gratification and quick, carefree returns on all investments leaves no time to wait for intricate things to be worked out, no time to evaluate true worth, no willingness to listen to those who contend with natural complexity. Knowledge earned during long years of study of people and nature is being cast aside to make way for technologies and financial systems that must satisfy their own, not human, needs.

Another example: a few years before Oved Shifriss left Rutgers, Professor Fred Hough, one of the world's great apple breeders, told me about the trouble he was having with the administration of the Experiment Station. "They want me to start submitting progress reports every six months! I try to explain that I'm breeding *trees*, not radishes—things don't change in a six-month time frame. But they don't get it."

Dr. Hough, too, had a grand plan. "I can breed an apple tree with good-tasting apples that will be largely resistant to the common pests and diseases. It won't require all these chemical sprays. I can put my hands on every bit of the genetic material I need. It will take a research group of nine people—that's not much for a university of this size. But all they care about is growing a bigger, shinier Red Delicious." Dr. Hough retired in disgust without embarking on his great project. When I last heard of him years ago, he was consulting on fruit tree breeding for Romania and Brazil.

Not only do we reject the gifts of genius present, we also spurn the offerings of genius past. Gene Logsdon, farmer and author, has written extensively about the brilliant methods our agricultural forebears worked out for using the gifts provided by nature, what modern ecologists would call "ecosystem services."[3] A choice example is keeping chickens and cows together in the same farmyard. The chickens pick the face flies and horseflies off of the trusting and cooperative cattle, giving the cows relief while gaining nourishing insect protein in their own diet. A different kind of service is provided by the dung beetle, which feeds her larvae on little balls of manure she buries in the ground. In two days, a pair of dung beetles can bury an entire cow pie, which if left on the surface would smother the grass and attract cattle parasites. Underground, the manure serves to nourish the soil and to produce more of the useful dung beetles. The service is free; the cost of rejecting it is not.

These gifts have been rendered irrelevant on most modern farms. Chickens are raised in crowded factory facilities—with no freedom to move, with beaks cut off so that they can't peck each other, with artificial illumination night and day—and they are fed a semi-liquid diet synthesized from grain products and additives. Meanwhile the cows are also packed into enclosures and feedlots; their pests and parasites are controlled by insecticidal sprays and chemicals added to their diets. The same chemicals, passed through in the manure, are probably responsible for the widespread decline of the dung beetles, which pretty much disappeared around 1950. And the manure, itself, is now a pollutant, not a nutrient.

My stories of rejected gifts have been taken from agriculture, but I could have found them in any walk of contemporary life, from architecture to engineering, from business to medicine to education. The life of contemporary civilization is like the fairy tale of the three little pigs and the big bad wolf run in reverse. Having started with a house of brick, we have moved, with great fanfare, to a house of straw; it is hard to concoct a happy ending for such a story.

We have brought this misfortune upon ourselves; there is no wolf, no malign outside force to blame. The earth is full of gifts, the perennial gifts of nature and the gifts of human creativity worked out over thousands of years of trial, error, and sacrifice. In our insistence that we must reinvent and manage the whole world from scratch, and do it quickly, in our confidence that money and power will replace, with something better, anything that once worked well, we have cut ourselves off from the bounty available to us.[4] But the money is dwindling and the power is waning. Soon, those who remember the gifts we have rejected will try to recover them. Will they be in time? Or, like my chance of reading to Dr. Williams, will the gifts be gone beyond recall?

---

## THE USES AND RISKS OF ADAPTATION

**adapt** *v.* **2.** *intr. & refl.* [to] become adjusted to new conditions.

*Oxford Modern English Dictionary*

When my wife Joan and I were newly married, we lived in a north Jersey suburb not far from the New York state line. Every weekday morning we drove down the Palisades Interstate Parkway to the George Washington Bridge and crossed the Hudson River to Manhattan, where I taught and Joan was a graduate student. The parkway runs along the Palisades, a magnificent, igneous bluff that flanks the west bank of the Hudson and faces, on the far shore, Yonkers, the Riverdale section of the Bronx, and Manhattan. Wooded parkland extends on both sides of the road for its entire length until just before the approach to the bridge, where many lanes of superhighway converge on the toll booths. We loved the woods along the parkway—they calmed us before our immersion in the chaotic city, and soothed us when we left it at the end of the day.

That was before we went on our honeymoon, a three-week hike on the Appalachian Trail (interspersed with some hitchhiking on country roads), from Springer Mountain, Georgia, to the border of the Great Smokies in North Carolina. The forest we walked through was a mixture of tall pines and an incredible variety of native hardwoods—an experience of natural diversity that was overwhelming. Nearly every tree we saw was new to us, yet we could feel the pattern and cohesiveness of the forest as a whole. Rhododendrons formed a closed canopy over our heads, fragmenting the June sunshine into a softly shifting mosaic of dappled patches. We stepped on a carpet of rhododendron petals.

The trip was over all too quickly. The plane carrying us back descended through a dense inversion layer of black smog before touching down on the runway at Newark. Home. We were depressed and silent. The ride from Newark Airport to our house took us on the Palisades Parkway. For the first time, we became aware that the woods along the parkway were dominated by thin, ungainly *Ailanthus*, with their coarse (and, we knew, rank-smelling) foliage, and by other weedy species such as the lanky *Paulownia*. Suddenly, these exotic species seemed very much out of place. The second-growth forest looked raw, monotonous, and badly assembled. We grew more depressed. It took about a month before the Appalachian memory faded and we became accustomed once again to the Palisades parkland. Then the *Ailanthus* disappeared into the leafy background, the flowers of the *Paulownia* gave a welcome splash of purple against the green, and the disrupted state of the forest strip was all but invisible to us after the days in Manhattan.

I have described a very benign and useful example of adaptation. Although I drive the Palisades Parkway infrequently now, I'm glad it is there and grateful to those who created it and maintain it. But what about our other adaptations, the majority of them, the thousands of small and large adjustments that we make to the changes that technology and crowding are imposing on us and on our surroundings?

My teacher, the great naturalist Archie Carr, once remarked (I think that he was quoting someone, perhaps Aldo Leopold), "Someday we will all learn to love gas stations." In a similar vein, while musing about the question "What will be pleasant in [the year] 3000?" he wrote in 1964 in the preface to his book *Ulendo*:

> Our race is frighteningly malleable. *Adaptable* we would rather call
> it. Already masses of people have been conditioned to processed

cheese, to bread that I think is made of plastic foam, and to gaudy fruits from which all taste has been jettisoned. . . . People can be made not just to live shoulder-to-shoulder in tiers, but to enjoy living that way.[1]

Orwell said pretty much the same thing in the final pages of *The Road to Wigan Pier*, first published in 1937. He attributed the worst adaptations in modern life to "the decay of taste" caused by our acceptance of and demand for more machine-made articles, and by our consequent distancing of ourselves from nature. "Mechanization," said Orwell, whatever its advantages, "is out of control."[2] If we extend Orwell's thoughts, taste—that outdated notion of an absolute standard of quality—may well have been our last cultural barrier against indiscriminate adaptation.

Adaptation is usually a healthy and necessary process, as the apostles of progress frequently observe, but passive, heedless adaptation is dangerous. Surely some adaptation is needed if we are to preserve our sanity in this rapidly changing chaos of our own making. People who live in large cities, for instance, have to make many behavioral adaptations, such as suppressing their natural eye contact with strangers, if they want to survive. There comes a point, however, when the hidden cost of adaptation gets too high. I can give examples, some of which I will expand on in later chapters.

We are adapting to enforced helplessness. We used to be able to fix things ourselves; now working parts are often sealed and can only be thrown away when broken, or require specialized tools to repair. At the same time, there has been a marked decrease in the durability of many products. We have adapted to this rapid, planned obsolescence with hardly a whimper.

We are adapting to less personal contact with others. Synthetic voice answering systems, optical scanners in high-speed checkout lines, automatic tellers, self-service gas pumps, "distance learning," computer-assisted work done in the isolation of home, and e-commerce all take the flesh-and-blood people out of our daily lives. We are getting used to doing without them.

We have long become adapted to television as the source of much of our news, entertainment, and companionship. In other words, we have replaced two-way communications involving our full participation with one-way reception that allows no effective response. At the same time, we have become adapted to the steady substitution of commercials for content in the programs we watch. Our tendency to adapt to a commercial environment is often cynically exploited, as is also the case with National Public Radio, whose management is steadily increasing the number and frequency of commercials on air time that was once promoted as public-supported and commercial-free.

Our adaptation to television has caused many college students to expect television-style lecture presentations, with showy electronic slides and a predominance of glitter over content. As has been pointed out by the Yale political scientist, statistician, and graphic designer Edward Tufte, "Slideware often reduces the analytical quality of presentations. In particular, the popular PowerPoint templates (ready-made designs) usually weaken verbal and spatial reasoning, and almost always corrupt statistical analysis." PowerPoint, claims Tufte, "leads to over-generalizations, imprecise statements, slogans, light-weight evidence, [and] abrupt and thinly argued claims."[3] Tufte is not the only one worried about our adaptation to this new form of information transfer. In August 2003 the Columbia Accident Investigation Board at NASA released the first volume of its report on the causes of the accident that destroyed the space shuttle *Columbia* in February 2003. One of the causes, they argued, was that NASA relied too much on presenting complex information via PowerPoint, rather than the usual printed technical reports. The PowerPoint presentation of the shuttle wing damage was confusing, with nested bullet points and irregular short forms that were difficult to decipher quickly. The board noted, "It is easy to understand how a senior manager might read this PowerPoint slide and not realize that it addresses a life-threatening situation."[4]

We have adapted to the convenience of E-Z Pass and other electronic toll systems, and in so doing have lost awareness of the amount of tolls that we actually pay. The economics columnist David Leonhardt, writing in the business section of the *New York Times*, said:

After an electronic system is put in place, tolls start rising sharply. Take two tollbooths that charge the same fee and are in a similar setting. . . . A decade after one of them gets electronic tolls, it will be about 30 percent more expensive on average than a similar tollbooth without it. There are [*sic*] no shortage of examples: the Golden Gate Bridge, the George Washington Bridge and the Tappan Zee Bridge, among them.

Electronic tolls are only a small part of the story. With 55 percent of consumer spending now done electronically (cash and checks divide the rest), people are no longer aware of how much they are spending. As Leonhardt states, "With electronic money, it's much easier to be carefree."[5]

We are adapting to less courtesy and consideration from others. This is partly because we are adapting to the imposed pressure of time on all our actions. Efficiency has become the only criterion of excellence. Courtesy takes time without yielding immediate benefits. Similarly, we are adapting to the constant demand for quick decisions. No longer is there time for reflection when we are faced with complex issues. Communities cannot survive without courtesy and time for reflection.

We are adapting to a relentless erosion of our privacy, the contraction of the boundaries of personal space in our business and social relations. The cell phone not only has intruded on the privacy of others in restaurants, meetings, buses and trains, and even the theater, but the users themselves have adapted to the fact that everyone within hearing distance will learn about their personal affairs.

We have become adapted to an indoor climate regulated so that it is the same during winter and summer.

We are adapting to the headlines that tell us that business executives are making many millions of dollars a year even as their companies lay off tens of thousands of workers.

We have adapted nicely to the conventional modern wisdom that there is no special reason for people to live near the place where they were raised and where the bulk of their family resides.

And we are adapting to the removal of nature from our lives and its replacement with symbolic substitutes. The nature special on television replaces the walk in the woods; the voyeurism of ecotourism replaces the joy and toil of the garden. The change of seasons, the migration of swallows, the shower of the Perseids, and the lamentable decline of the dung beetles go by unnoticed.

Although our easy adaptability is not a primary cause of our troubles, it is allowing damaging forces to do their worst with little hindrance. Our passivity to the manipulation of our culture by short-term, antisocial, economic interests is making us pushovers for destructive change. More serious is that our children are growing up to accept our adaptations as baseline, natural behaviors. They have internalized them.

What can we do to stop ourselves from adapting to conditions that should not be tolerated? There is a Human Rights Watch that reports on rights abuses around the world. Maybe we need an Adaptation Watch to warn us of dangerous lapses of taste and to remind us of truly self-interested, community-fostering behaviors that we are losing. We will have to speak out loudly, publicly, and often to point out the lethal changes in our environment and rouse the sleeping memories of what we were and the awareness of what we are fast becoming.

In the end, however, the war against dangerous adaptation is a personal struggle. Each of us must set his or her own limits of the tolerable—which changes to our world we will accept and which ones we will resist. I drive a car; I don't watch television. I occasionally send a fax; I sharply limit the number of e-mail messages I send and receive. These choices are not arbitrary, although they may be disputed. What matters is that they are *my* choices.

Becoming the arbiters of our own adaptations often means selecting a less popular and more inconvenient way of doing things. Enormous forces backed by vast resources push at us to adapt. But in resisting, we are not without power. We have the power of knowing that we can fight successfully without having to win the whole war, and the power of knowing that nature and community are on our side.

---

## WHEN MACHINES REPLACE PEOPLE

I used to enjoy listening to the National Weather Service forecasts on my shortwave weather radio. An endlessly repeated taped message updated every few hours might be less than thrilling, but the voices of the half-dozen or so forecasters made it come alive. Each one had an identifiable style and intonation; it was easy to assign personalities, even faces, to them.

In the '80s, the announcers were all men. There was the one I labeled the grand elder, with his pontifical voice and distinctive, rolling rhythms. When cost-cutting forced the station to move from Manhattan to the grounds of the Brookhaven National Laboratory, way out on Long Island, he disappeared from the airways. Perhaps the daily commute on the Long Island Expressway was too much for the old fellow. I am sure I wasn't the only listener to mourn the loss of his avuncular cadences.

Another announcer who appealed to me spoke fluently until he came to an American Indian place name such as Manasquan or Wanaque (both in New Jersey). Then he hesitated. I could imagine the look of terror in his eyes when he scanned the next line of the script, and there it was, a word with fearsome Q-sounds or daunting combinations of consonants and vowels. If I had had any way of getting in touch with him, I would have comforted him by explaining how lucky he was to be broadcasting in the New York–New Jersey metropolitan area. Up in northern Maine, the forecasters have to cope with names such as Caucomgomoc and Chemquasabamticook.

Some announcers proclaimed their individuality with what seemed like deliberately odd pronunciations of common words. The most original was the fellow who figured out a new way to say "climate," an achievement I would have thought was impossible. He did it by lengthening the separation between the two syllables and heavily stressing the second: "cly-*matt*."

Eventually, the Weather Service hired its first woman announcer, a welcome addition; she made her mark immediately by shortening the phrase "Here are the latest Central Park observations" to "Here is the latest Central Park." This bothered the linguistic purist in me, but when all the other announcers took up the abbreviated phrase, I saw the wisdom of her choice; by simply eliminating one word and changing another, she had instantly broken through the glass ceiling and become a member of the club.

The only announcer who could keep me listening attentively through several cycles of the tape came on the air sometime in 1998. She was obviously a very young woman with—a first for the weather station—an unmistakable Hispanic accent. In the beginning, I was irritated because she spoke quickly and I couldn't catch all the words. But there was a quality of sweet and wistful earnestness about her voice that made me suspend judgment: I decided to give her a chance. Over the next few weeks I noticed that she was slowing down and was becoming perfectly under-standable. Maybe I'm getting used to her accent, I thought. But I later learned that the weather service broadcast staff, in response to a flood of angry letters (some viciously racist), had been working hard with this young Puerto Rican intern, coaching her on pace and pronunciation. Soon I found myself tuning into the station even when I knew what the weather was going to be, just for the sheer pleasure of listening to her.

Who would have dreamed that a weather forecast could be so charming, could evoke images of lost youth and vanished innocence, could whisper promise of a better world to come? Her voice did that for me, and when she pronounced the name of my state "New Yersey," I could hear it through her ears, against a background of surf on a Caribbean beach, with a gentle breeze rustling the fronds of the sway-ing palms. Of all the announcers, she was the one I liked most.

Then on Tuesday evening, February 16, 1999, everything changed. I switched on the weather radio and there was a new voice, a man's, deep, also accented, and strange. The rhythm of the speech was Scan-dinavian, I believed, but when I recalled the voices of Swedes and Norwegians I had known, they seemed different from this one. Every phrase and sentence was a distinct unit, with a pause of exactly the same length wherever there must have been a comma in the script, and a slightly longer pause after every period. Within phrases, the words were often run together, and some words were slurred, as if the an-nouncer were drunk. Words came in predictable bursts, utterly devoid of emotional content. I could not imagine the announcer's face, but his voice was grim. It was also intensely boring. I heard the tape twice with-out taking in what he said. I never did find out what the weather was going to be.

The next day I tuned in expecting one of the morning announcers, but there he was again, deadly, inhuman. Suddenly the penny dropped. The voice sounded inhuman because it *wasn't* human—I was listening to a computer! I felt as if I had received a blow. The day was ruined for me, regardless of the weather.

Why was I so upset by a fake voice on the weather station? That evening, my friends Mischa and Eugenia explained it to me. Mischa recalled the words of the Abkhazian writer, Fasil Iskander, who observed that in Abkhazia, where the farming and pastoral population is always outdoors and where the weather is a daily matter of life and death, nobody listens to the forecast on the radio. But in Moscow, with its sheltered apartments, office buildings, and public transportation, where the weather means much less in the lives of people, everybody listens. Listening to the weather report affords us an opportunity for nostalgia, says Iskander, a connection with a world of nature that most of us no longer experience in a vital way. There is no way that a computer-generated voice can inspire that kind of nostalgia.

A few days later, I spoke with someone at the National Weather Service, in Upton, New York. Yes, he said, it was a synthesized voice. No, nobody had been fired—yet. But yes, the point was eventually to save money through downsizing; that's what Jack Kelly, the assistant administrator in charge of the NWS in Washington, wanted. Yes, the voice was awful, but it would be improved. They were working feverishly on the software. Well, yes, there had been complaints. They expected them to die down in a month or two. That has been their experience with computer-generated voices elsewhere in the country: after a month, people stop complaining, they adapt.

I didn't want the complaints to die down, and I didn't want the voice to be improved. Even if they could get the thing to sound like James Earl Jones or Helen Mirren I still wanted it scrapped. Voice quality is not the issue, being human is, and that can't be synthesized either by computer science or biotechnology.

Most of us live a lifestyle that largely cuts us off from the weather. Now, we are not even allowed the nostalgic pleasure of listening to a live person—a person whose dreams and fears animate her voice—tell us about the sunshine or rain expected tomorrow. The mechanical announcer is an abomination, and I was angry at Mr. Kelly and the other administrators of the National Weather Service for imposing it on us.

Nearly a decade later, the electronic voices persist, and they haven't changed much. The woman's voice still says "pat-chief og" instead of patchy fog, while the men's voices are as boring and lifeless as ever. When I want the weather report I tune into a regular radio station. The weather reporters there are real humans. Every now and then, I try the weather radio again, hoping that the synthetic announcers have disappeared, but they never have. One thought consoles me, however: I can still go outdoors to see and feel what the weather is like. At least the synthetic announcers don't follow me there. And as for Mr. Kelly, or whomever has succeeded you, I have a friendly warning: In the downsized, lonely, and unnatural world you've helped to synthesize, you may find that nobody gives a damn about you anymore.[1]

---

## PSEUDOCOMMUNITIES

To begin with, a little anecdote. There is a classroom, an ugly, badly shaped, windowless room in a modern university building designed with students not in mind. In this room there is a small class, my class. We have rearranged tables and chairs in a semicircle around my place to defy the terrible ambience and to allow all twenty-five students to see and hear each other and me. Class is in session; I am talking. Two students sitting together in the front row—a thirty-year-old man with a pager on his belt and a twenty-year-old woman—are speaking to each other and laughing quietly; they see that I am looking at them, and they continue to laugh, not furtively or offensively but openly and engagingly, as if I weren't there. I don't know what they are laughing about. Both of these students will eventually receive an A in the course for their exceptionally fine work.

As I speak to the class, a part of my mind is thinking about these students and wondering why they are laughing. Is there chalk on my face? Is my fly open? Have I repeated myself or unconsciously misused a word or inverted a phrase? Then I remember something Bruce Wilshire wrote, and the paranoia fades. The laughter has nothing to do with me. Bruce, a philosopher who works in a nearby building on campus, described the same attitude in his own classes in his book, *The Moral Collapse of the University*. This passive and casual rudeness, a fairly new phenomenon, has a simple cause, he said. "I sometimes see students looking at me as if they thought I could not see them, as if I were just somebody on their screen."[1] When my students were laughing, it didn't occur to them that I would be bothered—at that moment they were treating me as if I were a talking head on television.

That's what it is; I am sure of it. The students (at least most of them) and I inhabit different worlds, even when we sit in the same room. In my world, the people I speak with are real; if I offend them they are hurt and angry *with me*, if I give them pleasure they smile *at me*, if I bore them they find an excuse to move away *from me*. We are alive to each other.

The world of my students is far more complex, a hybrid world, a world in transition. For them, some of the old world survives, but it is confusingly intermingled with the new, artificial world of electronic communication. Up to now, this has been mostly one-way communication; the recipients are physically and mentally passive. The faces on the television screen speak and speak—my students have been watching and listening in some cases for five hours a day or more, since the age of one, two, or three. I cannot compete with this. Three hours of class a week for fourteen weeks is no time at all compared with the television they have seen. Their authority figures are two-dimensional and these figures cannot hear. They neither take offense nor do they rebuke; their brief utterances are well suited to the wandering, superficial mentality fostered by the ever-flickering monitor.

The world of television, by inducing passivity and unresponsiveness, has cut many of the human threads and connections that once bound people together into working communities. Lewis Mumford compared life in front of the television screen to life in a space capsule, frightening in its absolute isolation.[2] Passivity and alienation are not communal virtues.

To the extent that television has weakened American communal life, it has weakened communal power, the only effective power to limit consumption, pollution, and the degradation of nature. This would appear to lock us into a hopeless spiral of decline, for television promotes these very evils while weakening the communal ability to resist them.

If this were all the threat that electronic communication has to offer, we might conceivably find ways to cope with the challenge, formidable as it is. We might, for example, take advantage of the fact that television, which has helped to ruin communities, has not eliminated the desire of most people to be part of a community. A lonely consumer, however passive, is a dissatisfied consumer—unstable, even rebellious. Could we use this instability to fashion a revolt against television? Not any more. Not in the age of e-mail and the Internet.

Perhaps unconsciously, the developers of electronic communication have come up with the perfect counterstrategy to prevent us from using the loneliness of people caught in the television culture to wean them away from the tube and back into the community. This counterstrategy is the creation of *pseudocommunities*, my word for assemblages of electronically linked people. Pseudocommunities have arisen as substitutes for the real ones that are going or gone. Pseudocommunities are making everyone (or nearly everyone) feel good again, are replacing enervating passivity with a semblance of activity, creativity, and choice; but they are keeping the reality of true neighborly, communal responsibility and judgment far away.

Among the most familiar, albeit relatively primitive, systems of electronic, two-way communication, precursors of the pseudocommunity, are the recorded or electronically voice-simulated phone operators. Few people are left in the United States and Canada who have not had their blood pressure raised five or ten points by a patronizing recorded voice saying, "If you want to discuss your bill, press one now; if you want to speak to a nurse to schedule an appointment, press two now; if this is an emergency, press three now; if. . . ." Making fun of this is like shooting a sitting duck; I leave it to those who earn a living as nightclub comedians or humor columnists. But a few words will not be amiss. These systems are a threat to communities, paving the way for pseudocommunities, in that they accustom us to dealing with facsimiles of people in our daily lives. Whether people or facsimiles do a better job is irrelevant—there is more to life than maximizing the efficiency of daily transactions (although electronic operators rarely do that—they waste vast amounts of time). Daily transactions between real people are one of the things that can make life worth living.

There is usually nothing that one can do about recorded and simulated voices except hang up, which is not always practical. The ruse of pretending that I don't have a touch-tone phone has been nullified by the demand that I speak the number of my response to the electronic voice decoder. Only in the case of the artificial information operator at 411 is there any remaining possibility of satisfaction. When the bright, phony voice says, "What city please?" I answer "wuszch wuszch wuszch" in a low monotone. This baffles the computer and a real operator picks up. How the phone company will deal with this in the future I don't know, but they will find a way. Perhaps a sublethal zap of electricity through the receiver's earpiece will modify my Luddite behavior.

Lately, when the wonders of the age of information and communication have got me down, I have revived my spirits by rereading three exceptional books, Jane Austen's *Emma* and Elizabeth Gaskell's *Wives and Daughters* and *North and South.*[3] These books take place almost entirely within a few miles of the houses of the central characters, and they describe the incredible subtlety and wealth of interactions, for good and evil, that one experiences in a real community. In the sort of communities described by Austen and Gaskell, the passions and activities of love, hatred, sexuality, compassion, selfishness, and intellectual intercourse, modulated by and expressed through the life of the community, take on a complexity and richness that cannot occur in an electronically facilitated pseudocommunity. In a later chapter, I will give Jane Austen some of the attention she deserves, but here I must get back to pseudocommunities.

It seems that almost every advance of our technology brings more social disintegration. Consider interactive video communication, which is already a part of distance learning and many telephones. To the voice in the receiver has been added the face on the screen, and this changes everything. The feature that kept the conventional, hardwired telephone from destroying communities is the *lack* of visual information that accompanied the voice. The disembodied voice was a constant reminder of what the telephone really brings about: communication between people who are actually, demonstrably, perhaps distressingly distant from one another. Like a letter, a phone call was received in private from someone who was elsewhere. Add a picture, and the privacy and sense of distance are disturbed, replaced by an illusion of proximity, a mockery of context. This is another step on the road to pseudocommunity.

But the danger to communities of interactive video is trivial compared with that of e-mail and the Internet. Here I fly in the face of conventional opinion; this technology is supposed to be a liberating force in society, and in some ways it is. Using electronic communications, one can send a message to any person in the world who is in the network, or direct it to large groups of people simultaneously. It is much harder now for repressive governments or powerful interests to hide news that runs counter to their interests. What better way to promote the free democratic exchange of ideas, to create a "global community"? But there are several catches.

First, the centralized nodes through which all Internet messages pass will always make them vulnerable to censorship and tampering. And compared with old-fashioned spying, covert electronic manipulations are faster, cheaper, and pose little personal risk for the people carrying them out. The technology of privacy and security of communication is in a perpetual race with the technology of invasion and manipulation of the information transmitted. Like the arms race or the evolutionary battle between plants and the insects that eat them, there are no permanent winners and losers.

Far more significant is that the Internet fosters the sensation of being part of a community of people living in the same region and working, creating, and playing together for the common good. But the sensation is only that, for at the end of the day when you in Vermont and your electronic correspondents in western Texas, Delhi, and Yorkshire go to sleep, your climates will still be different, your time zones will still be different, your landscapes and soils will still be different, your local environmental problems will still be different, your cultures and histories will be different, and, what is most important, your neighbors will still be different; and while you have been creating the global community you will have been neglecting them.

The speed and simplicity of communicating electronically can be alluring and habit-forming. Unlike ordinary letter writing, anything that comes to mind can be conveyed instantly with little bother, and can receive an instantaneous response. Thousands and millions of streams of consciousness are accessible by cable or wireless. But this is not necessarily an advantage; it is often a problem. In a proper, durable relationship, many thoughts, after careful reflection, should be left unsaid. Careful reflection takes time and sometimes privacy, assets that we have stupidly wished away. Equally important, if we hope to make valid judgments about things and people, we must have information from all the senses, information that can never be conveyed fully by words, or even pictures on a monitor. The failure of many electronically formed love affairs, once the couple finally meet face-to-face, is a case in point. There is no easy, glamorous way to be part of a community. The phrase "global community" is an oxymoron.

Pseudocommunities detract from the real work of community building, which, although deeply gratifying, requires painstaking, persistent efforts and perpetual learning that continue as long as one keeps on breathing.

Communication is good and necessary, but not at the expense of communal integrity, which requires a balancing measure of separation of one community from another, and, at times, of one individual from another. Electronic communications systems lack this balance, this subtle regulation of communal function. In the pseudocommunity of e-mail and the Internet it is becoming harder and harder to maintain the kind of personal boundaries that add strength and diversity to real communities and keep them, in most cases, from flying apart.

Constantly there are new products and systems of electronic communications coming on the market. Reality is being replaced with virtual reality. Where will this end? For it will end in the not-too-far-off future. It will end because the global pseudocommunity is and will increasingly become economically unstable. As energy and material resources rapidly grow scarcer and more expensive, the Internet, which is very resource-consumptive, may not be as readily available as it is now. Moreover, for the great majority of users, electronic communications do not help to create real, durable wealth or benefits. The human and natural resources devoted to these electronic systems are not resulting in a socially acceptable outpouring of *necessary* material goods and services. Instead, production, local economic stability, and communal security are sacrificed to transient efficiencies, destabilizing luxuries, and the quick profits of distant entrepreneurs.

Although they do not create much real wealth, the new communications enable wealth to be shifted rapidly from place to place. Under these circumstances of global free trade, instantaneous global finance, instantaneous global exchange or theft of wealth-producing ideas, and facilitated global exploitation of distant resources, it is becoming difficult for most real communities—and people—to continue the slow accumulation or even maintenance of assets that is a condition of survival. As our real communities and nations become more impoverished, the accumulated wealth and social order that have supported the information research, the electronic hardware and software, and the enormous energy consumption of the Internet[4] will be beyond our means to provide. At this point, the whole network is likely to fragment and contract.

The new systems of communication will also be rejected for social reasons. In addition to being exploitative and expensive, consumers rather than generators of wealth, pseudocommunities are thin, transient, and above all unsatisfying.[5] Although it is fun to play electronic games simultaneously with hundreds of partners in several dozen countries, this kind of fun does not sustain any but the shallowest of existences. Moreover, the loss of real human contact, combined with the breakdown of defined boundaries of self and community, will not be tolerated by most of us forever. Already, research is indicating that those who spend the most time on the Internet and in e-communication are among the loneliest people in our society.[6] Pseudocommunities are seductive, but at some point, most of us will rediscover that face-to-face friends and coworkers are superior to virtual ones.

In an article in the *New York Times* about campus e-mail, reporter Trip Gabriel wrote that on many campuses, electronic communication is preventing the development of meaningful, communal relationships.

Dormitory lounges are being carved up for clusters of computers, student unions are declining as gathering places, and computer-wired dorm rooms are becoming, in some cases, high-tech caves. . . . James Banning, an environmental psychologist at Colorado State University who surveyed some 100 university housing officers last year, remarked: "Universities are saying: 'Oh, my God, they're in their rooms. How can we ever build a sense of community . . . if they don't come out?' "[7]

In one extreme case, a student described by Gabriel communicated with his roommates by e-mail even though they were sitting a few feet apart in the same room. But another student, who had become dissatisfied with electronic socializing, said, "It's easier to just meet someone. You learn how much of a difference it makes to see someone in person and actually talk to them."

Electronic forms of communications are still relatively new and exciting. Indeed, for some who are infirm and handicapped, electronic communication can provide a life-saving source of human contact; and for anyone the Internet, if used judiciously, is a wonderful and quick source of information. We should not, however, confound the value of the Internet for providing information with its value as a substitute for community.

I remember a photograph of a baby monkey in a psychological experiment; it was being raised in isolation with a surrogate "mother" made of wire covered with terry cloth.[8] The monkey was clinging to the device, but it looked profoundly sad and anxious. For most of us, in the end, as the dust gathers and the glamor fades, our pseudocommunities of silicon and plastic and liquid crystal will prove no more comforting and no more nurturing than a surrogate mother of wire and terry cloth.

---

## OBSOLESCENCE

At the end of the Cretaceous period, the last dinosaurs disappeared from the earth, setting off an evolutionary jubilee among the Milquetoast-like mammals that survived them, and preparing the ground for what was to become, sixty-five million years later, a permanent source of gainful occupation for scientists whose job it is to wonder why the dinosaurs died out. Scores of reasons have been given for this remarkable concatenation of extinctions. Global climate and sea level were changed by a city-sized asteroid striking the earth near what is now the Yucatán, or by a massive set of volcanic eruptions, or by the solar system passing through the core of a giant molecular cloud, perhaps colliding with a supercomet loosened from the Oort cluster, which orbits the Sun beyond Pluto. Theories of catastrophic extinction abound. Some of the most daring even conjure up the specter of an unseen companion star to our Sun, named Nemesis, whose eccentric orbit brings a wave of potentially deadly comet showers—and extinctions—every twenty-six million years. But there are also paleontologists who argue that the dinosaurs went away gradually, not suddenly, over a period of millions of years, and that toward the end they coexisted with the earliest hoofed mammals, including ancestors of horses, cows, and sheep. If extinction was gradual, a different line of thought opens up: perhaps the dinosaurs died out because they couldn't adapt and compete in a changing world. The big lummoxes were obsolete.[1]

I first heard about the dinosaurs' obsolescence back in my student days. It was as satisfying a notion then as it is today, especially if you didn't think about it too hard. Here were these lumbering, pea-brained reptiles, barely able to walk and chew gum at the same time, while all around and underneath them, cleverly hiding behind clumps of primitive vegetation and cleverly burrowing in tunnels in the ground, were the nerdy but smart little mammals about to emerge from the shadows and begin their ascent to glory—somewhat, it occurs to me now, like Bill Gates in the waning days of American heavy manufacturing.

Unfortunately, if one clutters up this elegant picture with a few more facts, the outlines begin to get blurry. Some of the dinosaurs were small and presumably agile, possibly warm-blooded, and maybe even feathered. They had much better eyes than the mammals had then or have today, if we can judge by the dinosaurs' surviving relatives, the modern reptiles and the birds. (This is not to say that the mammals didn't *start* with good eyes, inherited from their own lines of reptilian ancestors. But they lost them during endless eons of subterranean and nocturnal skulking, and when they came out into the sunshine again they re-evolved, almost from scratch, a jerry-built substitute.) As for cleverness, if brain size is a key to evolutionary survival, what happened to the Neanderthals, whose brains were bigger than ours? Why did they become extinct? Finally, although it seems almost churlish to point it out, consider the alligators and the crocodiles. We could call them obsolete: they are big, they have small brains, they are slow and ungainly on land, and they are distant relatives of dinosaurs. They are also still around. And with a little protection from overhunting, they can do surprisingly well in our unforgiving world.

Actually, I haven't the faintest idea why there are no more dinosaurs, and if you are reading this to find out, you can stop now. The questions that interest me more are, where did the idea of obsolescence come from, and why has it become so popular? Does its insistent presence in our lives help or hurt us? Is it possible that our constant fear of becoming obsolete is leading us along the very road that the dinosaurs traveled?

Some sense of obsolescence is, I suppose, an inevitable by-product of developing technologies. Foot soldiers in biblical times learned the hard way about obsolescence when they first did battle against men in iron chariots. George Sturt, working in his wheelwright's shop in Victorian England, knew all about obsolescence, the "immensity of changes" in machine technology that instantly "separated employers from employed . . . robb[ing] the latter of the sustaining delights which material used to afford to them."[2] But seldom if ever in history has obsolescence become such a leitmotiv in our lives as it has in this electronic age. Anything that is gone, even the dinosaurs, is now described as obsolete.

The social and economic costs of obsolescence are incalculable: perhaps the most important is our frightening loss of access to older information, as fragile digital storage media deteriorate and, more important, as changes in software and hardware render older documents inaccessible. As Jeff Rothenberg, senior computer scientist at the Rand Corporation, pointed out: "Shakespeare's first printed edition of [his sonnets] exemplifies the longevity of the printed page: the words are legible after almost four centuries. . . . But digital media can become unreadable within a decade." As Rothenberg notes:

> Information technology continually creates new schemes, which often abandon their predecessors instead of subsuming them. . . . [In the future], if we need to view a complex document as its author viewed it, we have little choice but to run the software that generated it. What chance will my grandchildren have of finding that software 50 years from now? . . . Storing a copy of the operating system on the CD may help, but the computer hardware required to run it will have long since become obsolete. . . . It is only slightly facetious to say that digital information lasts forever— or five years, whichever comes first.[3]

Standardized relational database systems can be devised, but Rothenberg shows that they have serious flaws that can render translation difficult or impossible. The comparatively short functional lifespan of digital storage media introduces another complication. And even if all of these problems could somehow be overcome, what chance is there that they would be implemented by the software and computer hardware industry? For declared obsolescence accompanied by constant change, whether needed or not, is a proven way to keep one's product ahead of similar competitors.

Obsolescence today implies much more than the simple recognition of what is new. It demands the absolute rejection of the old, as if the new were only validated and confirmed by the denigration of everything that came before it. To call some object, process, idea, or person obsolete is to render it beneath contempt, incapable of improvement. What force can have shaped such a destructive attitude, which at a single stroke cuts us off from our past and devalues our soon-to-become-obsolete present? One force powerful enough comes to mind: the drive of corporations to maintain and increase profits, a drive empowered by advertising. As General Motors' Alfred Sloan and his successors in many industries have long known, obsolescence sells products. The fear of becoming obsolete is being carefully taught at this moment to countless consumers around the world, drummed home by all the energy of the media and reinforced by every branch of our education industry, from kindergarten to graduate school. This is a catch-22 situation, because there is nothing wrong with profit-making companies trying to make more profits—in fact, we expect them to. And there is nothing illegal or even indecent about declaring one's older products obsolete while trotting out the new ones. Yet the end result of this process is destructive to everyone. "Let them eat the future," Wendell Berry once wrote.[4] How many of us are left, I wonder, who share with Berry the precious knowledge that the future is inedible?

Selling the creed of obsolescence is hard work because obsolescence is an unnatural concept, especially when applied on a time scale of only months, years, or even centuries. By unnatural, I mean that it does not really describe what happens in nature. Consider the American chestnuts, just one hundred years ago the dominant tree of the northeastern hardwood forests, which are now teetering on the edge of extinction. They were brought low in a decade by an introduced fungal parasite—did that make them obsolete? Extinction is often the result of a rare and highly specific event in the life of a species; the passing of that species is not a general indictment of its quality. The idea of obsolescence, as currently employed, is too vague and value-laden to be of use in biology, and it certainly doesn't apply to a diverse group of animals such as the dinosaurs.

Would we have been so quick to describe the dinosaurs of the late Cretaceous era as outdated if we did not live in such an obsolescence-conscious age? The dinosaurs no longer exist, hence the dinosaurs were obsolete; they were losers. What a strange thought, considering that more than 99 percent of all the species that have ever lived are now extinct. It makes about as much sense as saying that baroque music is obsolete, somehow inferior to the classical style that evolved from it, which itself became obsolete at the dawn of the romantic period.

Does it do any harm to apply the unnatural idea of obsolescence to nature? Probably. I am not worried about the dinosaurs. They are gone. But what about the frogs, the rhinoceroses, and the less conspicuous species of endangered nature? How tempting to write them off as obsolete (especially rhinos)—losers unable to go with the flow. Not only does this attitude undermine the conservation of vanishing species, but it distorts our perception of our own place in nature. We pretend that we are no longer part of the grand scheme but are sitting on the sidelines judging the participants. It is a fine show for a little while, yet in reality we have been given neither permission nor power to remove ourselves from the parade. We will be judged along with the rest, by criteria that are certain to be more complex than anything we can understand or predict, and nothing like our simplistic notions of winning, losing, and becoming obsolete.

Think about *Limulus*, the horseshoe crab, that curious marine invertebrate whose over-fished remnants are even now, as I write, clambering awkwardly out of the sea along the Jersey beaches, taking advantage of the spring tides to leave their eggs in the sand in the darkness of a new moon. If we knew them only as fossils in the Burgess Shale, we might say, "What preposterous creatures. How inefficiently they must have moved, with those ponderous shells and thin legs, dragging that grotesque spike. They were born obsolete, an evolutionary dead end. No wonder they are gone."[5] But the horseshoe crab is still here, looking much as it did fifty and one hundred million years ago, still bearing a strong family resemblance to its eurypterid ancestors who have not swum the seas these four hundred million years.

Horseshoe crabs and alligators teach us that the word *obsolete* comes much too easily to our tongues. Even within the limited context of our own technology, we should be careful where we paste that label. Did it work? Did it last or create the conditions for its own regeneration? Was it beautiful? Did it give pleasure? Was it a critical link in a larger process? Did it *need* to be replaced? Is its replacement an improvement? Is it possible that it will be needed for our survival in the days ahead? Slap! On goes the label—*obsolete*—and none of these questions matter, none are asked. To call something obsolete boasts an omniscience we do not possess, a reckless disregard for the deep currents of history and biology, and a supremely dangerous refusal to look at the lasting scars our technology is gashing across our planet and our souls.

---

## ACCELERATING SOCIAL EVOLUTION

The threat or promise of change brings out the frail nature of mankind's psyche. And sudden change is an imposition of instability. The rational argument, from its modern beginnings, has tried to avoid dealing with this reality. The multitude of abstract social models—mathematical, scientific, mechanical, and market based—are all based on an optimistic assumption that a schematic reorganization of society will be good for the human race. . . . [H]uman beings do not respond effectively to this sort of manipulation.

John Ralston Saul, *Voltaire's Bastards*[1]

There is a battle of evolution going on that is quite separate from the one that makes the newspaper headlines. It is an invisible, unpublicized struggle, and its outcome may affect our lives and civilization more directly than the original evolution controversy ever did. The new protagonists are not science and traditional religion; instead, they are the apostles of the religion of progress versus those surviving groups and individuals committed to slow social evolution as a way of life.

To understand this other struggle, it is necessary to look at evolution in a broad context that transcends biology. Not just a way of explaining how the camel got her hump or how the elephant got his trunk, the idea of evolution can also be applied to the writing of a play that "evolves" in the mind of the playwright or the "evolution" of treaties, banking systems, and anything else that changes over time in a nonrandom direction. More to the point, it applies—in nonbiological ways that Darwin inspired but probably never dreamed of—to the evolution of relationships among people in organizations or communities, one of the most important evolutionary forces affecting our lives, and the most poorly understood.

In the past, the concept of evolution has often been misapplied for political purposes. Social Darwinism—roughly explained as the belief that the rich and powerful deserve their success because they are more fit than the poor and weak—is the most glaring, dangerous, and scientifically foolish example. Nevertheless, there is a legitimate application of evolution to social processes: namely the evolution of institutional and communal relationships, which is clearly analogous to biological evolution. Formerly, in a business, university, or community that remained undisturbed for a number of years, new connections and relationships among organized groups and individuals formed slowly, the accumulated result of countless small decisions and chance events, similar to the mutations and genetic recombination at the heart of much biological change. If one of the connections or relationships was grossly dysfunctional, it was usually eliminated by fiat or by common consent, in the same way that natural selection acts on an individual organism. And like biological evolution, which has resulted in a great many false starts, stagnant lines, and extinctions, the evolutionary pace and direction of evolving institutions and communities have often been inadequate to keep them from extinction, too. Evolution—biological and social—is not a kindly process, but overall it works; some species, some institutions, some communities always manage to survive and prosper if evolution is allowed to run its course. Without social evolution, it seems likely that no human systems could work at all.

In our time, strong and ruthless antievolutionary forces are at play. Social evolution is not being permitted to run its normal course. These forces, however different they may be, have a common effect: frequent, sudden disruption of established relationships, customs, and behaviors among people who work and live together. Cataclysmic changes—from ice ages to the impacts of giant meteorites—have disrupted biological evolution in former ages of the earth, wiping out vast numbers of species in mass extinctions. Now, human activities on a global scale are causing another mass extinction of species, and, in a terrible extinction of a different sort, the evolving social order of our relationships, institutions, and communities is being disrupted by the powerful, inhuman forces of modern technology and economics. Changes occur too rapidly, too extensively, and too often to permit social evolution to work properly; increasingly it breaks down altogether. The ability of our social institutions to develop the enduring human cooperation necessary for survival is being threatened. Evolution requires a critical blend of change and stability: change to provide new possibilities for coping with an ever-altered environment, stability to ensure that successful adaptations, whether new or old, are not immediately swept away and discarded in a continuous, blind, senseless upheaval.

Incessant reorganization is the most violent of the anti-evolutionary forces now loosed upon the industrial world and much of the Third World as well. Once an infrequent event, a last resort to be called upon for a system that had patently failed, reorganization has become both a routine management tool and an incidental consequence of the mergers and takeovers that dominate today's business. As Scott Adams, cartoonist of the anti-evolutionary chaos and author of *The Dilbert Principle*, has written, "People hate change, and with good reason. Change makes us stupider, relatively speaking. Change adds new information to the universe; information that we don't know. . . . On the other hand, change is good for the people who are causing the change. They understand the new information."[2] Necessary change is one thing; change as a way of controlling others is something quite different.

Reorganization is a mighty weapon in the hands of an insecure manager (a surprising number of managers seem insecure). It breaks up comfortable working relationships that might, if left undisturbed, allow employees the time and peace of mind to question and evaluate the managers' decisions. When reorganization strikes, hardly anyone is secure; personal connections are severed; paranoia increases; innovation and creativity decline. It may take months or years for the chaos to subside, perhaps just in time for the next reorganization. Frequent reorganization slows productive work and reduces its quality, both of which depend on stable, evolved relationships. Imagine, for instance, a jazz band or string quartet that has played together for years, each player attuned to the slightest nuances in tempo and dynamics of the others; then, suddenly, the members of the group are forced to switch instruments at unpredictable intervals and change places with other musicians waiting offstage. What would the music sound like under these conditions?

An unusually severe bout of reorganization hit my university a few years ago. A university committee empowered by the office of the provost and led by a distinguished professor in the social sciences terrorized the five campuses in New Brunswick for months, gathering "information" and threatening major reorganization. Eventually a report was delivered to the provost. We waited anxiously, not knowing what it said, devising strategies of response to the scenarios we imagined. It was like living under the wavering ax of a drunken executioner. Then the blow fell. Without warning, the provost's office itself was abolished by the president. The reorganizers were reorganized. Naturally, the committee's report vanished without a trace, and all of our strategies were totally inadequate to meet the new disruptions. A few weeks later, I called one of the administrators who had survived in the wreckage of the former provost's office. "I'm keeping a low profile," he confided quietly, "so that I don't get my head chopped off. Those of us left here move cautiously because we don't want to slip in the pools of blood on the floor." Since this reorganization, there have been several more reorganizations, each of them backed by specious rationalizations, and each causing totally unnecessary damage, while the remaining people who actually do the work struggle on as best they can. In the latest reorganization, for example, the names of two of our colleges were eliminated while their administrative structures were changed, causing disorganization and problems for students, and infuriating loyal alumnae of both institutions for no legitimate purpose that could not have been achieved in a less disruptive way.

The rapidity of the electronic transactions that have replaced older ways of doing business is another force that opposes evolution. Social evolution is slow, not as slow as biological evolution, but a lot slower than the frenetic, sound-bitten, microsecond-conscious blur that we exist in now. It takes a long time to get to know people, to work with them effectively and smoothly—a long time to learn what they can and can't do, will or won't do, what turns them on and what turns them off, when they can be trusted and when they can't be trusted, when to smile and when to frown. In an electronic world, there is little time for social evolution, which remains stubbornly resistant to any efforts to make it move faster.

One of the first electronic disruptors of social interactions was the fax. Ordinary letters are slower than faxes. There are times, especially in the case of routine communications that face a deadline, when a fax seems the only way to get the job done. But the speed of a fax has serious drawbacks; when I send a fax, I need to ask myself whether I can afford to work so quickly. Having the time for reflection is not a luxury and is not dispensable. A decision that seemed unambiguous when it was made may appear quite different after a night's sleep. In such cases, a rapid exchange of faxes will help to promote bad choices and the social disruption that follows.

Far more intrusive than the fax, however, are the digital systems that now pervade the workplace and manage our relationships with our friends and colleagues. Although billed as efficient conveniences, they are often anything but. There is no time to make things work smoothly, even if they once did. We are too busy learning how the new system works and how to get the information to and from the fellow in the next office. The hype and euphoria that surrounds technological change helps to mask the social and economic disruption that these electronic systems facilitate. Computer expert Paul De Palma states:

> Many studies, including some done by the National Research Council and by Morgan Stanley ... fail to indicate any correlation between productivity growth and information technology expenditures. Distressingly, the opposite appears to be true. As Thomas Landauer has pointed out in *The Trouble with Computers*, those industries that invested most heavily in information technology,

with the exception of communications, seem to have the most sluggish productivity growth rates. Though one still could argue that schools and colleges should continue to teach courses in microcomputer literacy . . . our time might be more profitably spent breaking the bad news to the public that pays the bills. In the process, we might also come to understand how a machine so patently clever as the microcomputer could have done the business world (outside of the computer industry itself) so little good.[3]

Just as technology-fostered speed (including the speed of introduction of new technologies) disrupts social evolution—the development of stable institutions and relationships—so does technology-fostered isolation. I have already discussed the isolating effect of e-mail and the Internet; the impact of this isolation on the fabric of institutions and communities can be devastating. It is hard to imagine how the process of social evolution can function when connections among people are limited to the narrowest of information channels and the workplace is fractured into a collection of individually staffed home workstations and far-flung employees communicating with each other by e-mail, text messaging, cell phones, and occasional video conferences. While this kind of atomized and shifting arrangement is most likely to be found in the service sector of business and in universities, manufacturing is also prone to technology-related upheavals. Factories are now far more portable or disposable than they once were. Increasingly, when their plant moves to a different state or country, workers have to choose between losing their jobs or abandoning their homes, friends, bowling clubs, music groups, favorite hiking trails, trustworthy auto mechanics, neighborhood shops, and children's schools to follow their workplace to its new location (if they have been offered that option).

Social evolution, like its biological counterpart, is a process of great depth and subtlety. Its best end products—lasting institutions, durable friendships, stable communities, accumulated wisdom, and gentle and productive cooperation—are among the highest achievements of our human existence. They help us live in a mutually beneficial way with the other products of biological evolution, which we collectively call nature, and they can shield us from nature's occasional violence and from the violence that wells up in every human society.

But the Second War of Evolution is upon us. Incessant disruption, superficial contacts, and personal isolation are making social evolution impossible. We are like brightly colored bits of glass inside a giant kaleidoscope. Oh, look at the beautiful pattern we make. Twist, it changes. Oh, look at this one. It's not so nice. Twist, it changes again. Twist. Twist. Don't look anymore, just hang on.

Better yet, let's get out of the kaleidoscope. Find others and slowly make our own patterns with them. Reaffirm quality over the false and deceptive claims of efficiency. As Erwin Chargaff, great biochemist and a father of molecular biology, once remarked: "There is no hurry; there never is any hurry."[4]

---

## WRITING

The semester was over, my grade rosters had been handed to the registrar, and I was home, writing sporadically and glancing through a book by P. G. Wodehouse, whose incomparable prose, as elegant as it is funny, is always very soothing.[1]

The thirty students in Conservation Ecology were very good that year—the year my frustration with their writing reached the boiling point. Most of them were informed, outdoors-loving, committed, hardworking people who cared for each other and seemed to enjoy the class. With a few exceptions, however, they couldn't write English, and I was still recovering from the effects of reading the term papers upon which their grades were largely based.

For years, I had been assigning a major term paper in lieu ("in loo," as one student wrote) of an exam. I used to get eighteen- to forty-page papers, acceptably or even nicely written; but by the mid-90s—and getting worse every year since—the students were struggling to reach fourteen pages with the help of triple spacing, margins you could drive a bus along, and type sizes usually reserved for the visually disabled. And the writing!

The first mistake the students are now making is to use the spell-checkers of their computer software as a substitute for proofreading. The results are papers in which all the words are spelled correctly, even if they are not the right words. Reading these spell-checked papers can be like trying to translate from Spanish to English based on the assumption that words that are spelled the same in the two languages have the same meaning. (This can lead to some confusion if, for example, the Spanish words are *sin, cabal,* or *saber.*) In one of the papers, I had to erase a long, marginal comment I had written when I realized that *illicit* was meant to be *elicit.* In another, I had trouble with the philosophical implications of the word *modal,* used throughout the text, until I turned it into *model,* which made more sense.

Then there were words such as *begum* ("a Muslim lady of high rank," my dictionary told me), which didn't seem to fit easily into the context of a paper on the genetics of coyotes, and which didn't have an obvious substitute. When I came across words like this, I just left them: it was easier to imagine Muslim ladies of high rank dealing with coyote genetics than it was to find the intended word beginning with "b" or, alternatively, ending with "m."

Complicating the difficulties brought on by the spell-checkers is a new system of placing commas, apostrophes, and other punctuation marks, a system whose secret logic has not been divulged to me. It may be a flaw in my training, but I find it difficult and terribly time-consuming to read a paper in, which, the comma's have, been placed in a seemingly, random, manner, and apostrophe's are used to make word's plural. If a student then further complicates the writing by the sentence structure and scrambling, and not only is it inverted and, because, of all of the, parenthetical phrase's, and especially using, noun's as adjective's like desert fish habitat loss (DFHL) with so many acronym's, maybe, five or, ten, to learn and forget for each, paper and with an organization based on the idea that if you, repeat, a sentence enough with a few small change's, each, time then you have organized your thought's so you see the affect I mean effect it has on me and why after reading the paper's I wanted to kick the dog except because of his teeth which are very, large. The paragraphing:

too.[2]

When I finished grading the last paper on that day when my frustration with the students' writing came to a head, the dog was asleep on my foot. Getting up quietly and carefully, I went outside to plant potatoes in the back garden with my son Sam. I made the furrow; Sam placed the potatoes at nine-inch intervals and covered them up with soil. His older brother Jon was disdainfully shooting baskets in front of the house, not interested in what we were doing, but Sam's adolescence took a different turn; it didn't seem to keep him from enjoying potato planting with his father—yet. By the time we finished, I was calm enough to go into the house and tell my wife about the papers.

Joan listened to my lament, echoed it with some acerbic comments about the term papers she was grading for her course on global ecosystems, and said, "You realize what's happening, don't you? They're writing the way they talk. They don't know the difference between written and spoken language—that's part of the problem."

"Why don't they know the difference?"

"Because they don't get practice writing anymore, except things like text messages. To make matters worse, they don't read either, so they don't get to see how other people write. That's why they don't know the difference between 'affect' and 'effect'; they sound the same when they're talking, which is all they know how to do."[3]

"That's true," I answered. "Most of my students are seniors, and they tell me this is the first term paper they have written. They've never written anything longer than three or four pages."

The next day, I was still thinking about it. What, I wondered, are they doing in elementary or high school while they are not writing? Well, the answer to that is pretty simple; they are being tested, or, more likely, being prepared for being tested. To survive in the modern world of American education, it is absolutely necessary to be able to do well on standardized tests. And no matter what the makers of these tests say, the best way to do well is to prepare early (third grade, for example) to learn the different forms of standardized questions, to memorize (but never use in daily life) standardized lists of words and definitions, and to work through many sample, standardized problems. Testing is, I believe, a necessary part of education, but when it *becomes* education, when it displaces the teaching of what it is supposed to test, something is seriously wrong. How, I mused, have the educational testers, thousands of educational psychologists, statisticians, computer scientists, and school administrators, managed to pull off this massive swindle? How have these "professionals" convinced us that so much testing is necessary and that these tests don't themselves interfere with and misdirect the process of education? Part of the answer must be our infatuation with anything that smacks of science and statistics. But there is more to it; the testing industry has done a wonderful job of hiding the assumptions that underlie its tests, some of the most dubious assumptions one can find in a democratic society.

The first assumption of standardized testing is that for any question there is nearly always a clear and evident right answer amidst a sea of wrong answers.

The second assumption is that complex ideas and subtle nuances of meaning can be reduced to brief questions with short answers that can be graded by a computer. Even written essays on these tests can be evaluated objectively and satisfactorily by poorly paid graders who do not know the writers, and whose work output is itself being timed by the management of the testing company.

The third assumption, one of the most dangerous, is that a test score will reveal the testee's present worth, promise as a future student, and his or her level of success in life itself, years later, after formal education is done.

There is a final assumption, the master assumption behind all the others. It is that written language can be abstracted from daily life without doing it any damage. What I mean is that the testers, professionals to the core, have helped to take written language away from my students, who are mere amateurs; they have separated it from life and have made it a disembodied part of what we mistakenly call the educational process. Considering the testing assumptions that the students have grown up with—that there are always evident right answers, that complexity reduces to simplicity, that objective tests describe the present and predict the future—no wonder they have trouble writing (and thinking) about the problems of the real world.

Writing gains strength from frequent practice, but only when inspired by the events of daily living.[4] For instance, in a short story, I came across a description of someone as "weedy" in appearance. Looking this up in my dictionary, I find that it means "tall and thin; lanky and weak-looking; of poor physique." But as a gardener who knows weeds from experience, I can give the definition added meaning. True, weeds are often tall, thin, and unattractive; they are also fast-growing, persistent, and, for all of their poor physique, surprisingly tough. The people who first used "weedy" as an adjective must have known these things—they were putting part of their lives into the word, thus making their language a bridge between nature and mind.

The writing of most of my students' term papers shows that the bridge is out; the reality of their language is a static and contrived virtual reality, consisting of disembodied images without true life, without sustaining passion. In no way does it reflect the world of their active, outdoor being. Even the punctuation and paragraphing are dead and disconnected. A comma should be nothing more than a place to take a quick breath without losing a train of thought; a dash is a deeper breath; a paragraph is a pause that allows you to look around at the landscape and see where you have been and where you are going. The apostrophe, most abused of all punctuation marks in the virtual world of my students, has two functions—one as an indication of something missing in a contraction, and second as an indicator of possession. But these vital functions are obscure to many students: the apostrophe is now only an archaic mark vaguely associated with the letter *s*. One tree, two tree's. When writing works properly, its very words and structure should relate us to our world—assuming that we have a real world to relate to.

If written language is no longer an organic outgrowth of people's lives and environments, how can it be used to communicate the urgent problems of those lives and environments and how can it help in the search for practical and lasting solutions? It won't do to say that we will simply become a postliterate, oral/electronic culture where written communication is merely rapid speech made visible, as in text messages. Writing can be and should be more than frozen, spur-of-the-moment speech—it is thought enhanced by time and care, and connected to nature and history. It is the essence of human permanence.[5]

But it does not do to dwell on these things for too long a stretch. So if you will excuse me, I need to get away to my garden, as I did on that first day of overwhelming discontent with my students' writing. There are no potatoes to be planted, but I have to see whether the chicken wire fencing is keeping the rabbit away from my carrots and radicchio.

# THREE

---

## TOWARD A SUSTAINABLE
## ECONOMICS

The way to . . . deliver ourselves from our own destructive-
ness, is to quit using our technological capability as the reference point and standard of our economic life. We will instead have to measure our economy by the health of the
ecosystems and human communities where we do our work.

Wendell Berry, Life Is a Miracle[1]

## AFFLUENCE AND AUSTERITY

A funny thing happened to me on the road to affluence. I have talked about it with others often enough to discover that it is a common happening and an important one. But before you read any further you should know that what you are about to encounter is presented as a middle-class problem—nobody dies in an unheated garret, nobody is thrown out on the street with their meager belongings, nobody is in prison as a birthright, nobody is caught in a drug runner's cross fire. For all its bourgeois setting, however, my story is about the aspirations of our whole society; directly or indirectly it affects us all.

What happened is that as I got more affluent, I didn't feel any richer. Affluence, according to my *Oxford Shorter English Dictionary*, is an "abundance of worldly possessions, wealth." Setting aside quibbles about what constitutes abundance, there is no doubt that I now have more money, property, and other worldly possessions than I did, say, twenty-five years ago—especially if books count as worldly possessions. So why don't I feel wealthy? I can best explain by starting at what was for me the beginning.

I grew up in a nice house in one of the better residential neighborhoods of Passaic, New Jersey, a city composed mostly of immigrants and their children and grandchildren. The town of Clifton, a more prosperous community, began one block over; you could tell you were leaving Passaic because there was a sign. (It was not until I went to camp in Maine that I realized that in other parts of the country there can be open space between towns.) Our house had wood siding, two stories plus an attic and basement, three bedrooms, and a garage in back. It had cost my parents $6,600 in the late 1930s—a good deal of money for a struggling young doctor who was helping to support his bankrupted parents.

The property, narrow and moderately deep, amounted to about one-sixth of an acre. For the first eight or ten years of my life, there was a vacant lot next door. I remembered patches of milkweed inhabited by big, tiger-striped caterpillars that turned into monarch butterflies, even in a jar. I also remember my feeling of loss when the milkweed was torn up and replaced by an ugly yellow house. I could walk my affectionate but dim-witted beagle to the park nearby where there was a pond with a waterwheel at the outlet, and, once, a dozen newly hatched, brownish-black snapping turtles scuttling across the grass to the water.

Whenever I stayed home sick from school, I got to see the daily events of the household. Time never dragged; there was always somebody coming to the door. The side door, next to the kitchen, got the most traffic. I made a fort out of the heavy, white-painted, hardwood kitchen chairs and watched the arrivals from my protected vantage point. First was the milkman from Sisco Dairies, wanting to know how many bottles of pasteurized, unhomogenized whole milk to deliver, and whether we needed any light or heavy cream. Then the truck from Dugan Bakery pulled up: did we need a coffee cake, hard rolls with poppy seeds, or Danish pastry? The soda man brought cases of seltzer water, black cherry, and the fluorescent orange-colored soda to which I was addicted. In the late morning or early afternoon, Mr. Cohen, the fruit and vegetable peddler, showed up.

"What's good today?" my mother would ask.

"I got beautiful broccoli—and cheap," he answered.

My mother looked suspicious. "All right, let me see it." Then Mr. Cohen would walk to his truck and bring back a head of broccoli and maybe a cauliflower (which I hated). Haggling would begin; there were more trips back and forth to the truck; eventually, fruit, vegetables, and money would change hands.

In my early years, the iceman would arrive every few days, carrying a big block of ice with his wonderful, hooked tongs. By the time I was in third or fourth grade, a gas refrigerator replaced the icebox, and the iceman stopped coming. That a gas flame should cool things off seemed unnatural to me. It still does.

Sometimes the side doorbell would ring, and a shabbily dressed man with a stubble of beard would stand at the door.

"How do you do, ma'am," he would say to my mother, and maybe I would get a nod.

"Just a minute and I'll fix you something," my mother answered. Then she would make a hefty sandwich of whatever leftovers we had from the previous night's dinner—I remember pot roast—and would pour a glass of milk. The man sat outside on the side steps to eat, and later knocked on the door to return the plate and glass. I noticed an etiquette to these encounters: my mother never asked any questions, the men were always polite and appreciative without being demonstrative, and they expected to eat outside sitting on the side porch steps. Tramps were always traveling. I don't think I ever saw the same one twice.

The front door was used by the postman, who dropped the mail through a little slot with a hinged, elaborately carved iron cover. He came around ten o'clock and again, for the afternoon delivery, at three. When I was old enough to write letters, I learned that first-class mail to other parts of northern New Jersey took one or two days to arrive. Other than the postman, the front door saw a stream of visitors—aunts, uncles, and friends who came just to talk or to consult my mother about the various community service organizations that she helped to direct. Outside in the street there was still more activity. Every few weeks, on different schedules, the junk man, the rag man, and the knife-and-scissors-sharpening man would drive slowly by ringing their bells, pausing now and then so that people could have time to come out of their houses and call to them.

If my mother had an errand or appointment that took her away from home when I was sick, Josephine, who did the cleaning and helped with the cooking, could take care of me and tell stories about working in factories and about rural life in North Carolina. Josephine received a high salary for those days, and my father gave free medical care for herself and her family, as well as any medicines and X-rays they needed. My mother occasionally complained to my father about the salary, but he saw a great many patients a day at $2 for an office visit, and we could afford the expense.

On some days, my father would take me with him to his office, where I would read comic books and talk with the patients. If it was a Sunday when the office was closed, I could watch my father make out the bills. On some, he would write "discount" and reduce the bill by a third or a half or more. "Why did you do that?" I asked. "Because they can't afford to pay more," my father replied.

Why tell about these things? I guess it has been long brewing, but what set it off was watching a woman make a phone call. As I was driving through New Brunswick I saw her on a street corner holding a telephone receiver, trailing a cord, to her ear. At first I thought it must be a call on a police or fire emergency phone, the kind attached to utility poles; then I saw, next to her, a small plastic wind screen with the telephone company logo on it. This little open box was passing itself off as a phone booth! What happened to the real booths where you could go inside and close the folding door? A biological analogy popped into my head: the phone booth has become vestigial, like the human appendix, a victim of the cell phone. Yes, I know that the new "booths" have been around for a long time, but I had adapted to them unconsciously, and only a sudden quirk of memory made me aware of the substitution. The point is that our affluent society can no longer afford to provide real outdoor phone booths, with doors, to shield callers from wind, traffic noise, and garlic-eating eavesdroppers. True, those who have cell phones can go into their cars and use them there, but not everybody has a cell phone, and not everybody has a car.

I began to think about some of the other things we cannot afford. Two mail deliveries a day (soon we will be down to five or fewer per week), and ordinary first-class mail that arrives the day after it is sent. Furniture made out of solid wood, not chips glued together. Useful services that people need, where they live. Affordable helpers in the home who can work with dignity and for fair salaries and benefits (although exploitation is hardly a modern invention). Medical care individually designed with each patient's total well-being in mind. A house to live in when one is starting a family and doesn't have a big income. The time for casual visits with friends. The peace of mind and security to open one's kitchen door to a hungry stranger.

Some features of contemporary life are wonderful improvements over the way things were when I was a child. There is, for example, more justice in today's America, even if there is still not enough, and there is a new understanding of inequality. Tolerance of alternative lifestyles is growing rapidly, although not without resistance. Nostalgia can be a deadly trap if it is nothing more than a by-product of aging, but nostalgia is useful and necessary when it reminds us of better ways of living that are gone. I would not want us to return to the world of my childhood even if we could. I do want us to reconsider, in the full light of our experience, how we are going to define affluence in the world we will live in tomorrow.

I think about this when I am standing in a giant box store heaped with digital cameras, flat screen television sets, mobile phones that can do everything but fix a flat tire, video games, GPS units, and iPods, but with no salespeople in sight. These material goods do not make us affluent; they are a smoke screen to hide our increasing isolation and exploitation by an economy dominated by corporate forces outside our local communities.[1]

But the smoke screen is dissolving; soon, fewer of us will be able to afford even the palliative glitter that is much of what the world's industry and technology are able to supply. A country whose material goods are all produced by very poor people in other lands is in a perilously precarious condition. When I was a child, I remember hearing on the radio the postwar British speaking of austerity and the end of empire. They were very honest about it; even then their frankness impressed me. Now austerity is nearly upon America, too, spurred on by international trade, which is helping us export our jobs and capital; by a "high" technology that gives us inventions without regard for their final cost and consequences; by the dwindling supply of the cheap oil that supports all this; and by the staggering cost of the wars that protect the oil. Unlike the British, we don't speak publicly about austerity. It is coming just the same.[2]

What will Americans do when austerity exposes our latent poverty? The rage of a country packed with the adult equivalent of spoiled, frightened children is not pleasant to imagine. I expect denial, terminal indulgence, and random, often self-destructive violence. I hope also for something better. America is a tough society: innovative, resilient, and still full of humor. There are many people, old and young, who recognize true affluence and who are already spreading its seeds throughout our communities.[3] Soon they will have their chance to demonstrate the richness that comes from a simpler, more responsible way of life.

---

## ENERGY AND FRIENDLY FIRE

There is a critical issue in the energy debate that has not received the attention it deserves: the problem of "friendly fire." Friendly fire is the euphemism our military and press use to sugarcoat the presumably accidental killing of soldiers in battle by their own comrades.

The first time I saw the term *friendly fire* used in a nonmilitary context was in the book *The Argumentative Indian*, by the Nobel laureate economist Amartya Sen. Sen's point was that sometimes the institutions created by government to reduce inequities—for example, to bring more food to the hungry—actually make the situation worse. This he calls "friendly fire." I extend the use of this term: when the very success of an innovation or technology has side effects that cause the whole system to fail—when success brings about collapse—this is also friendly fire.

What does this have to do with the problems, especially the environmental problems, related to America's energy crisis? It becomes clear if we look at the possible ways of dealing with the energy crisis. I group them in two categories.

The first category includes all *technological* solutions. There is no need to describe them in detail: nuclear power, coal liquification, biofuels, hydrogen/fuel cells, efficiency gains in transforming, transporting, and using energy, nonbiological renewable energy, and others. Each has major advantages and serious limitations. Even the most benign have side effects. Many produce little more, or even less, energy than they consume in energy inputs.[1] And most have the same drawback: they are much more expensive than sticking a pipe in the ground and letting the oil flow out of a tap. Nevertheless, it is clear that some combination of these technologies will allow us to stretch our energy supply farther than current practice allows.

The second category of proposed solutions to the energy crisis includes the various methods of *conservation* of energy based on a more restrained lifestyle. It means, for Americans, consuming less and reusing more.

America has opted primarily for the first category: technology. It's easy to see why. Our present economy is geared to constantly increasing consumption, and dependence on goods and services we no longer provide for ourselves. There is a deadly combination of a sense that we are entitled to all these goods and services, and a fear that we need them, we can't survive without them. We don't worry about the ultimate cost, because we haven't the faintest idea what it is. In fact, we act as if there will be no cost. Thus, in America few of those advocating the new energy technologies are suggesting any reduction in overall energy consumption. Indeed, the opposite is likely to be true—continued low prices and new inventions encourage us to use still more energy. But there are two hitches: First, even taken together, all of the new energy technologies will probably not be up to the job of replacing cheap oil in running a high-consumption, high-waste society—although they will be critically important in running a responsible-consumption, low-waste society.

The second hitch is more serious, and here is where the friendly fire idea comes into play. If we use the gains from our new energy technologies to continue to increase our consumption and waste, we will find ourselves in a vicious spiral that decreases resources and increases environmental damage, *even as our energy technology improves*. Mathis Wackernagel and William Rees have explained this problem well:

> Unlimited cheap energy could simply expand human activities further, depleting other natural capital stocks. . . . To the extent that [energy technologies] contribute to increased consumption and accelerated resource depletion, efficiency gains [add] to the burden on the planet's limited carrying capacity.[2]

The central idea expressed here is about as controversial and hard to grasp as the statement that two plus two equals four. Yet a combination of techno-euphoria and denial has prevented most Americans from hearing the message.

Marine fisheries are a good example of this kind of friendly fire. Fossil fuels are the major energy inputs to the world's fishing industry. In the year 2000, 50 billion liters of fuel, mostly diesel, were burned to land a little more than 80 million metric tons of marine fish and invertebrates—this amounted to 1.2 percent of global oil consumption, about the same as used by the Netherlands, with an annual emission of 130 million tons of $CO_2$ into the atmosphere. The energy content of the fuel is about 12.5 times the protein energy content of the catch. Fleets catching luxury species—shrimp, tuna, swordfish—for the American, Spanish, Japanese, and similar markets have the highest energy consumption: two thousand liters of fuel per ton of catch. But, the energy efficiency of the fishing fleet has been declining steadily over time, because every year the boats need to fish longer hours and deeper in offshore waters as the overexploited fish populations progressively fail. So even if we improve the efficiency of diesel and gasoline engines by 10 or 20 percent, which is conceivable, and if our new technologies make more cheap fuel available, energy consumption will continue to rise in most fisheries until one by one they collapse, like the North Atlantic cod fishery. Abundant energy in the absence of scientifically regulated fishing strategies is killing the fisheries.[3]

Similar stories come from global forestry—one of many examples is the use of energy-guzzling helicopters to assist in the logging of steep slopes. Our insatiable demand for timber and timber products results in soil erosion and permanent deforestation in mountainous areas. The new energy technologies, in the absence of other moderating influences, remove a potential restraint on our ability to carry out destructive forestry. More friendly fire.

There are additional connections between increasing energy efficiency and accelerating environmental and social damage. Global trade, which is heavily energy-dependent, transmits introduced diseases of humans, plants, and wildlife and promotes the costly spread of exotic species. The annual cost of fighting exotic species in the United States has been estimated by Cornell biologist David Pimentel and his colleagues to be approximately $120 billion.[4] And there are other side effects: For instance, the importation from Mexico into the United States of luxury foods such as fresh green peppers, tomatoes, and string beans in winter is subsidized by huge inputs of energy for industrial agriculture and transportation. The side effects, regardless of whether the energy is sustainably produced or not, include the poisoning of Mexican workers by pesticides, chemical and mechanical impacts on Mexican soils, and the destruction of native Mexican cultures as community economies are disrupted by introduced subsidized commodities and the national demand for cheap, landless labor.[5]

Here is another example of friendly fire: aquaculture is now providing a substantial amount of the fish consumed in the United States. The aquaculture facilities, themselves, require large amounts of energy to run, especially in the case of carnivorous fish such as trout and salmon, whose animal protein is supplied by fish meal made from smaller fish caught by the same global fleets mentioned earlier. Aquaculture causes many damaging side effects—including genetic pollution of native fish stocks by escaped fish, global spread of diseases such as infectious salmon anemia (which is caused by a virulent virus that has adapted itself to intensive aquaculture), the destruction of the marine seafloor by bottom trawling, and marine pollution by food and wastes from the farms.[6]

Ecotourism, which is also dependent on cheap energy for long-distance travel, can be a great boon to conservation. Yet ecotourism has its dark side, as discussed later in the chapter "Conservation for Profit." It has caused serious harm to endangered wildlife populations and fragile ecosystems, including marine species and habitats.

And most of these consumptive uses of energy release greenhouse gases that work to alter global climate, which is raising sea level and appears to be increasing the frequency and intensity of large storms. Positive feedbacks add to the problem. For example, global climate change, itself, is predicted to cause drying of northern peatlands in central and western Canada, resulting in the release of excess $CO_2$ from unsaturated peat. This would probably raise temperatures still further.[7]

As long as energy technology helps Americans keep their same consumptive lifestyle, we have friendly fire on a massive scale. If energy were the only limiting factor we faced, perhaps our total faith in a technological fix would have some justification. But there are many limiting factors acting within the highly interconnected system that controls all of our lives. Forests, soils, fresh water, climate, ecosystem balances, emerging diseases, the growing gulf between rich and poor, and other interacting environmental and social variables are all approaching critical limits. Even if we can maximize efficiency in production and energy use, that does not necessarily solve the other problems—some, including the plight of environment and biodiversity, will be made worse by the continued availability of cheap energy. Like most enablers, energy technology is not a villain. We need it desperately, but we need it in a new and saner context of living.

What then is likely to happen? Ideally, the advances being made in energy technology would be employed in the United States together with needed changes in lifestyle—lower consumption and waste.[8] This would have many positive environmental and social effects around the world, although it would admittedly cause at least short-term economic disruptions in our major suppliers of goods and services: China, India, and others. At the moment, a responsible change in lifestyle in the United States seems a number of years away, although economic circumstances must inevitably force it to happen. Above all, we have to remember that the guiding principles of our actions must be moral first, technological second. Nobody expresses this more clearly and eloquently than Vaclav Havel, the former president of the Czech Republic:

> Whenever I reflect on the problems of today's world, whether they concern the economy, society, culture, security, ecology, or civilization in general, I always end up confronting the moral question: what action is responsible or acceptable? . . . We must return again and again to the roots of human existence and consider our prospects in centuries to come. . . . Maybe it is no longer a matter of simply promoting energy-saving technologies, but chiefly of introducing ecologically clean technologies, of diversifying resources and of not relying on just one invention as a panacea. . . . Technological measures and regulations are important, but equally important is support for education, ecological training and ethics— a consciousness of the commonality of all living beings and an emphasis on shared responsibility.[9]

There are early signs that the needed changes are beginning—Slow Food, community supported agriculture, local currencies, and other innovative new ways of living that do less harm are starting to emerge.[10] We can only hope that they will prevail, allowing the best of our energy technologies to flourish without causing any more casualties from friendly fire.

## DURABLE GOODS

I have been reading about the Lenape. It started as a kind of penance. Every day I pick up the newspaper and learn about different ethnic peoples who have been driven from their homeland by some other ethnic peoples or have been simply exterminated in place, a more lasting way of getting rid of them. One morning, while eating oatmeal and skimming over the news of the latest genocide in the *New York Times*, a thought occurred to me: Aren't I lucky that this isn't happening in New Jersey? Not that it couldn't, but at the moment it seems quite unlikely. Why not? I wondered. Is it our democratic government that protects us? Our melting-pot society? Then another answer came to mind. New Jersey has had an ethnic cleansing already, a very comprehensive one—the destruction of the Lenape.

As a child I was taught in school about the Lenni Lenape Indians, who once lived in New Jersey and then didn't. The teaching never went much further than that; it was pretty superficial compared with what my children learned about Indians (but not about the Lenni Lenape) in elementary school. So I decided to remedy the defect in my education and bought the only book about the Lenni Lenape I could find, a scholarly paperback by Herbert C. Kraft called *The Lenape: Archaeology, History, and Ethnography*.[1] I bought it out of guilt—as I said, a penance. Not guilt for what was done to the native people of New Jersey, which wasn't my fault or even the fault of my ancestors, who arrived in America in the late nineteenth century, themselves victims of persecution. The guilt was because of a feeling that to live in the New Jersey–New York–eastern Pennsylvania region—that is, Lenapehoking—without knowing anything about its prior residents was disrespectful; it was a dishonoring, even an obliteration of their memory. Because of its scholarly nature, I expected my reading to be dull and penitential; surprisingly, the book turned out to be fascinating.

Names are important. I learned in Kraft's introduction that the name "Lenni Lenape" is as redundant as it sounds. (When I was in third grade I thought that "Lenni Lenape" was a silly name, and that prevented me from thinking of the tribe in a respectful way.) "Lenape," in the Unami dialect of the Delaware language, means something like "the real people," "the common people," or "the ordinary people." As Kraft points out, "Lenni Lenape" would mean "the common, ordinary people." "Lenape" is sufficient. I also noted that Kraft uses the term *Indian* without any apparent qualms. My few Indian acquaintances do also. One of them becomes angry when someone calls him a Native American; it is, he thinks, a euphemism favored by repressive government bureaucrats and their ilk.

But perhaps it doesn't matter what tribal name we use. After all, nobody, including the descendants of the original Indian residents, has the faintest idea what the various groups of Indians living in Lenapehoking called themselves before the arrival of the first Europeans. Even their languages have faded away. When Kraft wrote his book in 1986, there were only about ten people left who were fluent in either the Unami (southern) or Munsee (northern) languages, and all of those speakers were old. In 2007, there were none left—the language was extinct.[2] According to New Jersey historian John Quinn, William Penn, the founder of Pennsylvania, was an admirer of Munsee. In 1682, Penn described the language as "lofty, yet narrow, like the Hebrew. . . . One word serveth in the place of three, and the rest are supplied by the understanding of the hearer. I have made it my business to understand it, that I might not want an interpreter on any occasion."[3]

What, then, remains of the original Lenape, apart from a widely dispersed and variously acculturated group of people who can claim one or more ancestors who lived in Lenapehoking before Giovanni da Verrazano first sighted it in 1524? Most of the tangible remains, especially from the earlier and longer preagricultural phase of Lenape existence, are those tools of daily life that do not rot or disintegrate after centuries or millennia of burial in moist, mostly acidic soils. The list is shockingly short: spearpoints, scrapers, hammers, axes, drills, pestles, mortar stones, atlatl weights, bola stones, netsinkers, pendants, pipes, and a few other artifacts, all made of stone. Later on, there were arrow-heads, pottery, and a few kinds of implements and beads, some made of copper. That's about it.

The reason why I find this list shockingly short is because of the contrast with the durable artifacts of my own life. Not that I am excessively devoted to material objects, but almost everyone living today in an industrial nation is surrounded by the accumulated products of many centuries of human ingenuity, especially from this product-dominated century.

Without leaving the desk chair in my six-by-seven-foot home office (a converted closet), I listed, before I got bored, fifty different metal, glass, and plastic items that would probably survive burial in soil for a few thousand years. I cheated a little because I counted the pre-Columbian stone axe, spearpoint, and pestle that I have placed on my desk to provide inspiration. Other items on the list include a glass paperweight, a brass doorknob, a stained-glass butterfly made by Jon, a triaxial milligauss meter (for measuring electric fields like the ones under power lines), some computer disks, and the fountain pen that I am using to write these words. I haven't even looked in the places where most of my durable goods are: in the kitchen, bedrooms, living room, dining room, bathrooms, basement, or—it would take a week to total its contents—the garage. Just one room of my house holds more species of long-lasting objects than does the entire, pitiful collection of spearpoints and scrapers that is the visible legacy of the Lenape nation.

What does their paucity of durable goods say about the lives led by the Lenape? Does it mean that their existence was dull, instinctual, animal-like—on a lower plane than ours? Is it possible to be fully human and intellectually alive without wheels, pulleys, or writing implements? Does the mind's potential develop in an environment with such a meager and primitive set of fabricated objects? Many advocates of our own materialistic economics and technology would say no.

The landscape itself is silent on these questions. The Lenape left behind no pyramids, no giant stone effigies standing among the oaks and hickories on the Kittatinny Ridge overlooking the Delaware, no remains of passive solar housing aligned to the first rays of the summer solstice sunrise across the Raritan River. What they did leave behind, besides the names Kittatinny and Raritan, Kraft tells us, is the ghostly echo of a reputation, a whisper briefly noted by the self-centered Europeans who displaced and scattered them. The Lenape "enjoyed a noble status" among their neighbors—friends and enemies—before we came. The tribes around them, the Nanticoke, Conoy, Shawnee, and Mahican, "claimed a close connection" with them and called them the "grandfathers." Theirs was a social genius that resulted in achievements we moderns only experience in our utopian dreams. The Lenape had a "fiercely egalitarian" society characterized by no authoritarian leaders and a powerful sense of community. They cherished the environment that supported them, cared for orphans, and respected the elderly. And during most of the long years when they occupied the land, they appear to have lived in peace.

Since the glaciers melted off the face of northern Lenapehoking about ten thousand years ago, sea level has been rising, slowly at first, then faster. We believe that many of the early coastal settlements, fields, and hunting grounds of the Lenape were covered by the waves. An imperceptibly slow process at most times, the cumulative change would have been apparent during an unusually high storm tide, as the waters rolled farther inland than they had ever done in living memory. How did these prehistoric Indians respond to such catastrophic events? What did they think, what did they say as they stood on high ground under the night sky and listened to the sounds of the sea pounding at their frail huts, sapling frames covered with chestnut bark? Did they address the Great Spirit, Kishelëmukong, "He who creates us by his thoughts," in his place in the twelfth and highest heaven, with words like these?

> He who makes the Pleiades
>     and Orion,
> And turns deepest darkness
>     into morning,
> And darkens the day into
>     night;
> Who calls for the waters of
>     the sea,
> And pours them out upon
>     the face of the earth . . .

The words are those of Amos of Tekoa, a herdsman and tender of fig trees, first of the Hebrew prophets, who lived in what students of the Lenape have named the Early Woodland Period. I have no trouble thinking that similar words, albeit with different constellations cited, might have been uttered by some long-forgotten Lenape living at the same time among a people not yet swept up in the tide of what we call history.

It would be a great mistake to assume that the Lenape can be encompassed and defined by the durable goods they left behind. It is the sort of mistake made by a materialistic culture, a culture preoccupied with illusions, a culture confused about what is lasting and what is fleeting, confused about what is greater and what is lesser, a culture whose future may be no brighter now than that of the Lenape on that death-foreboding morning when the first Europeans stepped ashore on the beaches of Lenapehoking.

## PRESERVING OUR CAPITAL

A deeply felt aversion to spending accumulated capital is an ancient part of the heritage of most societies. Although my father was a doctor, not a businessman, he taught this to me. He had lived through the bankruptcy of his own parents during the Great Depression, watching as they gradually sacrificed the inventory of their store in Passaic, New Jersey, to keep the family in food and clothing. On one side of this store, my grandfather sold records, phonographs, and sewing machines and repaired the appliances that he sold; on the other side, my grandmother, a brilliant dress designer, prepared bridal gowns for customers who came from as far away as New York City. She would dress the brides on the wedding day, too, and was celebrated for her ability to make the plainest bride look beautiful. But as the depression wore on, business fell off, the customers stopped coming from New York, and the stock of goods dwindled away. There was no choice but to close the store; my grandparents' livelihood was gone forever. Later, my father and the oldest of his four brothers made it a priority to pay their parents' creditors in full, a decision that entailed sacrifices in itself. The lesson was handed down to me: you can spend your earned income and any interest you may have received, providing you first set aside a portion to increase your savings; but *never* spend the principal, your capital, except as an act of final desperation.

To most of us, capital is associated with business, yet the habit of preserving capital and handing it on to the next generation started, I am pretty sure, not as an economic or financial practice but as an agricultural one. In Neolithic societies, it may have begun when farming replaced hunting and gathering as the main source of food. From Anatolia to North Africa to Peru, the staple grains of wheat, rice, corn, millet, oats, barley, and rye, the legumes such as peas and beans, and other vegetables from squashes to radishes, were almost all annual crops. Every year, our ancestors needed new seeds and tubers for planting; without new seed they would have starved. So every year a portion of the harvest, including seeds from the finest and most productive plants, was saved for the next planting. This seed would have been considered inviolable; to eat it was to risk starvation and death. The idea of preserving capital, the saved-up goods and assets upon which future earnings depend, is a simple and inevitable extension of the ancient rule of saving a part of the harvest for use in planting next year. The story of Joseph in Genesis and the Koran is a parable that stresses this life-saving mandate to preserve capital. Foreseeing seven years of abundance to be followed by seven years of famine, Joseph received Pharaoh's permission to require that extra portions of food produced in Egypt during the years of plenty be saved in central storehouses against the lean years to come.[1]

In the past century, there were those who paid the ultimate price rather than consume the necessities of future generations. The Russian plant biologist Nikolai Vavilov, father of modern crop plant protection, organized and directed the gathering and classification of viable samples of crop seeds from all over the world in his Institute of Plant Industry in Leningrad (St. Petersburg). By 1941, as Vavilov, chief preserver of the world's agricultural diversity, was dying of malnutrition somewhere in the Siberian Gulag, a victim of Stalin's and Lysenko's hatred of evolutionary biologists, and as the German armies were approaching Leningrad, more than 187,000 varieties of wheat, rice, peas, and other crops from all over the world were represented in the institute's gene bank. Vavilov's deputies rushed to preserve the priceless collection, preparing duplicates of the most valuable samples, evacuating some to eastern Russia, and regrowing many of the older seeds on a small plot of land near the front lines. As starvation grew rampant in blockaded Leningrad, the institute's scientists did their best to protect the seeds from pillagers, rats, and exploding shells. Dimitry Ivanov, a rice specialist, died of starvation while preserving the several thousand packs of rice in his keeping. Georgy Kriyer, in charge of the medicinal plants; L. M. Rodina, a keeper of the oat collection; Alexander Stuchkin, a specialist in groundnuts; and other extraordinary men and women slowly starved to death rather than eat the agricultural capital that surrounded them. The continuing, heroic efforts of those who survived saved the gene bank for a largely oblivious posterity.[2]

Oblivious because saving capital went out of style after World War II. We have developed a flair for squandering that would be the envy of even the most accomplished wastrel. Show us something that our children and grandchildren will need for their survival, and we will find a way to spend it or ruin it, even if the spending and ruining require much effort. Some of these efforts are remarkably ingenious; take, for example, the draping of a massive government building or an entire island in petroleum-based plastic sheeting as—some call it—an artistic statement about light and form. Is one a grinch for asking how many barrels of our children's vanishing oil supply were used in this bizarre and short-lived public display of reckless waste? Is it far-fetched to see a similar heedlessness in the political effort to maintain the low federal tax on what is already some of the most dangerously underpriced gasoline in the world?[3]

One of the insidious causes of our pervasive thoughtlessness is that stocks of capital are no longer as easy to assess as they were when all we needed to know was how much seed was stored in the shed or how many sewing machines and phonographs were still on the shelves. Now, the car is in the driveway in Massachusetts or Montana, but the oil wells are in Kuwait, Nigeria, Saudi Arabia, and the Gulf of Mexico. Or other scarce commodities, such as helium, are usually so remote from our daily experience that most of us have no reason ever to give them a thought.

Our waste of capital is also related to the problem of exponential growth. When the practice of saving capital evolved, population and consumption were stable or were growing slowly by today's standards. In the second half of the twentieth century, a rapid increase in both occurred. This has confounded our efforts to save capital in three related ways that were described in 1978 by an American physicist, Albert A. Bartlett.[4] Bartlett applied a well-known formula concerning growth rates to the question of how rapidly resources are being used up. From this basic and simple equation, he developed principles that he thought every educated person should know and illustrated them with a series of vivid examples. First, if we assume an annual growth rate in consumption of 7 percent, the amount of a resource used during the next ten years will equal the total of all preceding consumption of that resource in history. Our conventional thinking, based on millennia of slow growth or no growth, does not prepare us for such a dramatic depletion of our resources. *Growth makes us underestimate our future needs for capital.*

Second, with exponential growth, the time at which the diminishing stocks of a resource still appear sufficient and the time at which the resource runs out occur surprisingly close together. Imagine, says Bartlett, a single bacterium put in a bottle at 11:00 P.M. The bacterium doubles every minute. By midnight, the bottle is full. When was the bottle only half full? The chilling answer: at one minute to midnight. By the time the more perceptive and activist bacteria first notice that they are running out of space and call a public meeting, it may already be too late to do anything about it. *Growth makes us exaggerate the time we have left in which to take action to save capital.*

And third, counter to all of our intuitions and to the claims of oil companies and politicians, in exponential growth the life span of a finite resource is only weakly related to the size of the remaining stock. In other words, the number of oil fields or copper deposits left to be discovered makes little difference in estimating the date when reserves of oil and copper will be exhausted. To return to Bartlett's bacteria, let us assume that at 11:59 P.M. an enterprising bacterium discovers three empty bottles off the north coast of Alaska. This quadruples the space resources available to the crowded population. The bacteria emit a collective sigh of relief. But how long will this new space last? The calculation is very simple—only two more minutes. How much oil is in the earth? It doesn't matter. Assume, said Bartlett, that the entire volume of the earth is oil and we can extract every drop. At the then current 7.04 percent yearly growth rate in oil consumption, an entire earth-full of oil would last just 342 years! *Growth makes a mockery of our hope that undiscovered resources of capital will save us.*

The list of dwindling capital resources is not confined to obvious commodities such as oil, coal, helium, and scarce metals; it includes items that we used to take for granted, including soil, fresh water, clean air, open space, trees, and fish in the sea. Moreover, there is another kind of capital than conventional natural resources, namely human capital, and we are wasting that, too. By human capital I mean the skills in agriculture, crafts, academic learning, mechanics, the arts, and all the other necessary human occupations. I also mean the community-fostering skills of judgment, patience, consideration, and knowledge. Human capital, like crop varieties but unlike iron ore in the earth or intact forests, does not preserve itself automatically until it is used or damaged. It demands an ongoing, active, proficient effort to prevent its inexorable decay. It can disappear in one generation—ten generations of work may not bring it back. Only a stable society can provide the continuity needed to preserve human capital.

Parents, teachers, and elders are the prime caretakers of human capital, and the low esteem that modern society accords to these civilization-preserving people helps explain why human capital is now melting away. In a world dominated by huge corporations and global trade, the need of humans and their societies for continuity has become subservient to a nonhuman economic demand for cheap production, with its ever-changing technologies and job definitions. In the present economic environment, stability is considered stagnation, and human capital is not considered at all.

In recent years, the job placement advisers have begun to say to graduating seniors, "You will each have to get used to the idea that you should not expect to have one job that lasts your lifetime—you will have to be flexible about what you do and where you live." In other words, stability is an undeserved luxury; traditions of wisdom and workmanship are not worth the effort to acquire, let alone pass on; skill is short-lived, or at best utterly generic; and saving and planning are the quaint activities of characters in history books and old novels. We have come a long way from the farmer who took comfort during the dark winter night in knowing that the seeds for spring planting were safely stored.

Today, even as "conservative values" are touted in the media every-where, the venerable custom of protecting our assets is rejected and vilified. Everything is topsy-turvy as self-proclaimed conservative capital-ists lead the assault against the conservation of capital. Here is an evil that no easy remedies will erase. Saving capital is hard work, and our quick-profit culture imposes extra social and economic penalties on those who undertake the effort. We are not being asked to make the extreme sacrifice of a Vavilov or an Ivanov, but pulling back from the seductive attractions of corporate culture—the supermalls, the television, the deadening conveniences, electronic and otherwise—that are eating away at our true and only wealth like a cancer, calls for a quiet heroism and a daring creativity of its own. Heroism in choosing the more diffi-cult and sometimes more expensive path; creativity in finding new ways of saving capital in an environment that despises prudence and has no concern for the future. As Bartlett has shown, we have precious little time in which to act, little time between our present relatively comfort-able awareness of the problem and the imminent moment when we will feel the full, crunching weight of a global shortage of capital the likes of which we have never experienced.

Again, I return to a wartime example of how to save our capital. In World War II, the leaders of the United States urged the public to save, and the public responded enthusiastically. As author Mike Davis reports:

> Lessing Rosenwald, the Chief of the Bureau of Industrial Con-servation, called on Americans "to change from an economy of waste—and this country has been notorious for waste—to an economy of conservation." . . . The most famous symbol of this wartime conservation ethos was the victory garden. . . . By 1943, beans and carrots were growing on the former White House lawn, and First Lady Eleanor Roosevelt and nearly 20 million other victory gardeners were producing 30 to 40 percent of the nation's vegetables—freeing the nation's farmers . . . to help feed Britain and Russia. . . . In Chicago, 400,000 schoolchildren . . . salvaged scrap for industry and cleared lots for gardens. . . . Gasoline was rationed [Eleanor Roosevelt set an example by accepting the lowest "nonessential" category of gas allotment—three to four

gallons per week].... the bicycle made a huge comeback....
[and] the Office of Civilian Defense (OCD) . . . encouraged "buying only for need" and set up consumer information centers that
gave advice on family nutrition, food conservation, and appliance
repair. . . . The OCD consumer committees challenged the sacred
values of mass consumption—the rapid turnover of styles, the
tyranny of fashion and advertising, [and] built-in obsolescence.

In the 1940s, Americans still remembered the Great Depression, and
many had grown up in an environment of frugal economy and saving.
This no doubt helped the wartime conservation efforts to achieve their
spectacular success. But the reforms and the spirit of conservation faded
after the end of the war, and the several generations that have occurred
since then have lived only in an environment of waste and consumption.
Are we capable of responding, once again, in such a public-spirited way
to the challenges that face us? Can we still, as Davis claims, "find
surprising inspirations and essential survival skills in that brief age of
victory gardens and happy hitchhikers"?[5] I think we have no choice but
to assume that we can.

## CONSERVATION FOR PROFIT

Since the fall of Soviet communism, the profit motive has been pre-
scribed more than ever as the cure for many ills, including the decline of
nature. Although a large part of the natural world has been damaged or
destroyed by unregulated commerce, various groups of conservationists
are trying to save some of the most spectacular remnants of nature,
both species and ecosystems—with more commerce. The idea of turning the tables and using the methods of exploiters to prevent more serious exploitation is an interesting one, but the risks are high and not
everyone in the business of saving by selling seems to have given much
thought to them.

I first came across commercial conservation in one of its early and
most dubious forms, the farming of sea turtles, a phoenix-like enterprise

that, despite its chronic inability to turn a profit, always seems to spring up renewed from the ashes of its last bankruptcy. Because of its biological and economic complexity, the farming of sea turtles nicely illustrates many of the problems inherent in the commercialization of conservation and is worth examining in some detail.

There are seven species of sea turtles in the world's oceans, ranging in size from the approximately one-hundred-pound Atlantic and Pacific ridleys and the hawksbills, to the huge leatherbacks, which can weigh as much as 1,200 pounds and possibly more. All except the loggerheads nest primarily on tropical or subtropical beaches, although some venture far into cold waters in between nestings. The eggs of every species are prized as food and in Latin America are considered an aphrodisiac. The leather is used for expensive shoes; the hawksbill provides shell for jewelry; and the hawksbill, loggerhead, and especially the green are taken for their meat, cartilage (for soup), and oil (for cosmetics). Indeed, the green turtle has been described as the world's most valuable reptile. Not surprisingly, nearly all sea turtle populations have been seriously depleted, and the great majority of the nesting populations that existed two hundred years ago are now extinct.[1]

The theory of sea turtle farming goes something like this: green turtles (and other sea turtles) are in trouble because nesting females and their eggs have been taken from the nesting beaches in huge numbers, the turtles are caught in the waters adjacent to the nesting beaches and at their distant feeding grounds, and in recent decades they have been drowned by the thousands in the trawls of shrimping vessels or snagged on the millions of hooks of the long-line fisheries. If unchecked, this depletion will lead soon to total extinction of these slow-growing animals. How much better, the argument runs, to remove an expendable percentage of eggs from the nesting beaches, hatch them, raise the hatchlings in safe environments to market size, and sell them. This will displace wild-caught animals from the market because of the easy, guaranteed availability of the farm-raised turtles and the higher, more uniform quality of the domestic product. As a bonus, we can set aside a few breeders to supply eggs in the future, thus creating a closed system with no net drain on wild populations. It sounds good, especially to those who know little about sea turtles and to those who have a vested interest in the operation.

From the beginning, sea turtle farming was strongly opposed by Archie Carr, the preeminent turtle biologist of the twentieth century, and by a sizable group of his students and colleagues, including me. The reasons we gave for this opposition, and which we still give thirty-five years later, are rooted in a mix of biology and economics. Here are a few of the more important ones.

The first problem is the unavoidable expense of coping with the biology of a nondomesticated species, a frequent worry for conservation entrepreneurs. Sea turtles mature slowly. Some green turtle populations may take thirty-five years or more to reach sexual maturity in the wild; they are not exactly broiler chickens. To speed this up, captive turtles must be fed expensive food enriched with animal protein, something akin to trout chow. The animal protein is primarily derived from fish meal made from marine fish. In the case of subadult green turtles, this changes them from vegetarians, normally eating turtle grass, to carnivores. Feeding efficiency is low because the pellets can be dissolved quickly or be washed away before the turtles can eat all of them.

Holding tanks are expensive to build and sea water is expensive to circulate and filter. Fencing off bays or inlets for impoundments instead of using tanks doesn't work very well as a rule; the fences get holes in them or are destroyed by storms, or the impoundments become heavily polluted. In both tanks and impoundments, unnaturally crowded turtles get sick from a variety of bacterial, fungal, parasitic, and viral pathogens. At the National Marine Fisheries Service's Galveston, Texas, facility for "head-starting" Atlantic (Kemp's) ridley turtles, twenty-seven major categories of disease were noted among hatchlings. At the Cayman Turtle Farm, on Grand Cayman Island, West Indies, flagship of the (originally) commercial operations, one outbreak of chlamydiosis among their green turtles is reported to have caused a die-off that cost the company hundreds of thousands of dollars. Other turtle-rearing facilities around the world have mortalities that typically range between 50 and 75 percent, and sometimes higher. Whether turtles released from these facilities spread disease to wild populations is not known, but remains a possibility.[2]

There also are problems concerning reproduction, the primary measure of the success of any conservation plan. Keeping a breeding stock of sea turtles for a large operation is economically (and biologically) impractical. I once calculated that it would take 293 turtles to supply just 50,000 eggs per year, even assuming that all biological obstacles to producing fertile eggs in captivity are overcome—an unwarranted assumption. The space required for these very large (150–350 pounds) breeders and the cost of feeding and maintaining them make this notion of breeding stock a manifest absurdity. Only the Cayman Turtle Farm has made a serious effort to supply its eggs from its own wild-caught and captive-reared breeders; the hatching rates from eggs laid by these breeders, averaged over a twenty-year period, was just 6 percent, compared with 70 to 90 percent at well-managed hatcheries at natural nesting beaches. Obviously something—whether diet, accelerated growth rate, or captivity itself—is terribly wrong.

So turtle farms are really turtle ranches, according to the accepted definition of the Convention on International Trade in Endangered Species (CITES). They are not true farms, because they are not self-maintaining—they can't produce all the eggs and hatchlings they need from their own captive-bred animals. To survive, green turtle ranches must take eggs from the dwindling wild populations every year. They refer to the eggs they take as "doomed" eggs, ones that would not hatch if left where they were laid on the beach—a conveniently slippery and subjective rationalization. As an added sop to conservationists, turtle ranchers have released, with much fanfare, some ranch-raised yearling turtles; a few hundred yearlings is supposedly an adequate replacement for thousands of eggs and hatchlings in the wild, most of which would, presumably, not have survived to maturity. Because of the complex life cycle and migrations of sea turtles, however, there is no evidence that the majority of released yearlings ever find and join their wild relatives, or that they produce viable offspring. After more than thirty years of releases of tens of thousands of head-started turtles of at least four species in a variety of programs, from 1996 to 1999, only nine head-started Kemp's ridleys were documented to nest in the wild (since 1999, more head-started turtles have nested, but hardly enough to restore a species).

The third, and most serious, problem is one of market stimulation, a condition that is common to most kinds of profit-making conservation. To sell their products and keep up the necessary cash flow, turtle ranches have to create a stable demand for turtle meat, soup, oil, and leather, and for dried, varnished yearlings, hatchlings preserved in lucite blocks, or any other turtle products people will buy. Because of the high costs of ranching sea turtles, all these products must be priced as luxury items. At the same time, the high prices and especially the developed, stable markets make it extremely attractive to poach wild turtles and find ways of laundering them into commerce. Poaching is almost always cheaper than ranching; therefore there may not be any way to protect an endangered species once the world market is stimulated, even in the name of conservation. The power of global demand erodes all safeguards. The problem is one of scale: the world market is just too big for most natural systems to satisfy. Thus the commercial ranching of green turtles inevitably brings us around again on the downward spiral, a little closer to the extinction of the remaining populations. By no stretch of the imagination is this conservation. Nor is it good business. After a series of owners and no history of profits, the Cayman Turtle Farm was taken over in 1983 by the Cayman Islands Government, I assume because of its value as an important tourist attraction (more than four hundred thousand visitors in 2006) and as a souce of jobs in the Cayman Islands, not primarily because of its sales of turtle products.

The problem of turtle "farms" extends to freshwater turtles as well. According to biologists Shi Haitao, James Parham, Michael Lau, and Chen Tien-Hsi:

> There are over 1000 turtle farms in China worth over a billion U.S. dollars. . . . We submit that these lucrative farming operations are a major threat to the survival of China's diverse turtle fauna. Turtle farms are a problem mainly because they are the primary purchasers of wild-caught turtles. Apart from increasing their total stock of adult animals, farmers are always seeking wild breeders because successive generations of farm-raised turtles show a marked decrease in reproductive capability. This reliance on wild-collected individuals indicates that turtle farming is not a sustainable practice. . . . The bleak future of Chinese turtles is mirrored in other commonly farmed wildlife such as crocodiles, snakes, and bears. . . . We predict that the gathering of these species into breeding facilities is an ephemeral phenomenon that will be replaced eventually by a permanent one: the extinction of wild populations.[3]

It is worth pointing out that the authors of this paper are, respec-
tively, from the Department of Biology of Hainan Normal University,
China, and the Chinese Academy of Sciences; the California Academy
of Sciences and the University of California, Berkeley; Kadoorie Farm
and Botanic Garden, Hong Kong, China; and the National Museum of
Marine Science and Technology, Taiwan. Their impressive and diverse
affiliations lend considerable weight to their claims.

In addition to the threat of wild animal farming to turtles, crocodiles,
snakes, and bears, we can add the farming of tigers. The largest tiger-
breeding facility in the world has been created at Hengdaohezi, in
China's Heilongjiang province. One hundred tigers a year are being
produced in what is nominally a conservation facility, with the stated
aim of reintroducing them into the wild. However, the facility seems
much more of a danger to wild tigers than a solution. In an editorial in
the journal *Nature*, the editors state:

> By taking no note of which animals breed with each other, [the
> facility] may be diluting the genetic stocks of the captive cats indis-
> criminately. Staff at the centre seem unclear on how to promote
> conservation, stating a desire, for example, to reintroduce the
> tigers into the wild without planning the measures that would
> enable them to survive. . . . Poor breeding management and
> detachment from the international conservation community has
> produced a group of docile animals that would stand little chance
> of survival in the wild. . . . China plans to reopen trade in tiger
> body parts for use in traditional Chinese medicine. Indeed, farm-
> ing the tigers for this use was the initial purpose of the Hengdao-
> hezi centre when the government opened it in 1986. . . . But most
> conservationists have condemned any repeal of the [CITES] ban
> on the tiger trade. Even if, as proponents suggest, a legal trade did
> reduce the price of body parts, prices would remain high enough
> to encourage poaching. Conferring legal status on the trade would,
> in the end, increase trading volumes and put more pressure on the
> wild population.[4]

The results of the inbreeding of the farm-raised tigers are further described in an article in the same issue of the journal:

[Geneticist] Liu Yutang . . . says physiological problems from in-breeding already run rampant. "Some tigers are very weak, have different stripes, high cub mortality rates, and reduced immune systems," he says. "And when they're all related, then it's easier for something to wipe them all out," he adds, citing the waves of deaths from bird flu and canine parvovirus.[5]

Tourists, three hundred thousand a year, love the Hengdaohezi tiger farm. People who understand and are concerned about the long-term impact of tiger-farming on the future of the species, may not share their affection for the operation.

Other difficulties exist when species conservation is promoted as a business venture, with or without the introduction of global marketing. Valerius Geist, an eminent Canadian wildlife biologist, has repeatedly warned against the dangers of selling wildlife or wildlife products. Among the many hazards he mentions is the spread of diseases by animals that escape from game ranches. These diseases include tuberculosis and spongiform encephalopathy (chronic wasting disease, related to mad cow disease and to Creutzfeldt-Jakob disease in humans), which he predicted would be transmitted to wild populations of buffalo, elk, and deer. Although it would be difficult to prove, the recent dramatic increase in spongiform encephalopathy, which by 2006 had spread to game animals in fourteen states in the United States and two Canadian provinces, may indicate that Geist was right.[6]

But responsible proponents of conservation marketing also exist. Gordon Grigg, an Australian zoologist and conservation biologist, has advocated the regulated market hunting of wild kangaroos as a profit-making incentive to sheep ranchers, allowing them to raise fewer sheep. Sheep, Grigg points out, compete with the native kangaroos for food and damage the sensitive vegetation of the Australian outback with their alien hard hooves and destructive grazing habits.[7] Crocodiles, alligators, iguanas, and other wild species have been ranched, with varied results. Because the species, environments, and human interactions are always different, each case has to be examined on its own merits.

The commercial exploitation of entire ecosystems, as opposed to single species, can again be very problematic as a conservation technique, for many of the same reasons. For years, the preservation of the

New Jersey Pine Barrens has been justified by some conservationists as a way of saving the enormous underlying aquifer of pure, fresh water. But if that water were ever to be exploited on a large scale, the water table would fall, and even a small drop in level would cause extensive changes in the ecosystem, including major loss of wetlands.

"Ecotourism" is also being widely advocated by conservationists, with incomplete knowledge of its effects in many cases. The problem of overuse of sensitive ecosystems is obvious—sensitive plants and soils react poorly to trampling, even by friendly feet, and endangered animals do not always thrive when watched by large numbers of friendly people in Land Rovers and tour boats. There is no question that whale-watching boats maneuvering next to whales, Land Rovers carrying ecotourists positioning themselves close to African wildlife, and large numbers of visitors to rain forests and deserts can have adverse effects on the species and habitats being visited. It has been shown, for example, that cheetah reproduction was depressed in heavily visited African parks, and mountain gorillas in Uganda have died from common colds brought by human visitors.[8] Coral reefs are damaged by skin-diving tourists.[9] Some negative effects are more subtle: it appears that ecotourism may have been responsible for one of the first recorded introductions of a primary human pathogen, *Mycobacterium tuberculosis*, into free-ranging populations of wild animals: banded mongooses (*Mungos mungo*) and suricates (*Suricata suricatta*) in Botswana.[10]

Similarly, the campaign to sell nuts and fruits from tropical rain forests, however well intentioned, has preceded any serious attempt to find out whether the exploited forests can sustain these worldwide sales. Botanists Rodolfo Vasquez and Alwyn Gentry reported that the collection of forest-harvested fruits by natives for sale only in regional markets, a flourishing trade, may not be sustainable in the highly diverse Amazonian forests around Iquitos, Peru. In these forests, there are hundreds of kinds of trees per acre, and fruit trees are widely scattered. There is evidence that there are simply not enough seeds left in the forest after fruit collection to replenish the populations of fruit trees.[11] On the other hand, the botanist Charles Peters and three of his colleagues who studied Brazilian and Peruvian Amazonian forests of low tree diversity, primarily swamp and flood-plain forests with only one or two dominant tree species, came to a different conclusion: these kinds of forests may be able to sustain "market-oriented extraction of fruits" indefinitely.[12]

Increasingly, there are highly publicized efforts to investigate and develop the commercial value of traditional medicines extracted from tropical rain forests. In one study, in Belize, botanists Michael Balick and Robert Mendelsohn found that the strictly local use of products gathered from small harvested clearings, with long intervals between cuttings, does not seem to damage the forest or the harvested species.[13] This is encouraging, yet it would be wrong to extrapolate from this finding and assume that no harm would come from commercial extraction of medicines on a grander scale. As in the case of sea turtle ranching, local forests may not be able to survive the demands of the world market.

Specific biological knowledge is needed to determine the feasibility of each venture in commercial conservation. On this biological basis, some projects, such as sea turtle ranching, ought to be ruled out from the start. Others, such as the use of forests to provide traditional medicines, may prove feasible in selected cases. Nevertheless, even in these latter cases the economic scale of the operation is likely to be limiting. There will be few schemes of commercial conservation that supply a globalized market without destroying the species and ecosystems they purport to save.

Finally, even if we do find an occasional example of conservation for profit that is not demonstrably destructive, many are; and real conservation benefits seem few and far between. Perhaps more important, we should ask ourselves whether saving nature by selling it is worth the moral cost of the effort. Stanford biologist Douglas J. McCauley, who argues that there is scant evidence that market-based conservation works, has written:

> To translate the intrinsic worth of nature into the language of economics . . . [is] akin to saying that civil-rights advocates would have been more effective if they provided economic justifications for racial integration. . . . We will make more progress in the long run by appealing to people's hearts rather than to their wallets. If we oversell the message that ecosystems are important because they provide services, we will have effectively sold out on nature.[14]

We are left with three questions: Can a sustainable human relationship with the environment be based on the premise that conservation must always pay for itself? Is there anything of lasting, intrinsic value left in this materialistic world? Can the world be preserved without rich people having to sacrifice some luxuries? I think the answers are clear.

---

## HOT SPOTS AND THE GLOBALIZATION OF CONSERVATION

Eight friends were seated around the dining room table, two bulging photo albums at the ready, about to start a round of that curious adult version of Show and Tell known as Our Summer Vacation.

Donna and Stanley had begun their album with pictures of Stanley delivering a lecture to clinical psychologists at the Catholic University of Campinas, in the state of São Paulo, Brazil. This was followed by page after page of lush, colorful South American scenery: snapshots of the mighty Iguazú Falls taken from the river below; towering walls of green vegetation along the Rio Paraná; a thick growth of native trees overtaking the exotic plantings left behind by the celebrated botanist Moises Bertoni at the remains of his agricultural experiment station in the Paraguayan province of Alto Paraná. Then there were pictures of Stanley and Donna with their Brazilian hosts, flanked by stately palms; pictures at their friends' beautiful beach house in Ubatuba; pictures of costumed samba dancers; and more pictures of scarlet and blue macaws, gaudy toucans, purple bougainvilleas, brilliant butterflies, and the rich magnificence of one of the last remaining patches of Brazil's Atlantic coastal forest.

The album provided a breathtaking display of natural and human exuberance painted in a profusion of vivid reds, blues, yellows, and, above all, greens. I could almost smell the myriad fragrances borne on the tropical breeze and hear one of my favorite sounds, the chattering of parrots far above in the treetops.

As the pages turned, the phrase "hot spots" came into my mind, that graphic term for those patches of exceptionally high biodiversity—many of them in the tropics—that receive the lion's share of attention from conservation biologists and others dedicated to the preservation of endangered species. I thought of the conservation argument that says if we want to preserve life, look first at the places where, acre for acre, the most kinds of life exist—where an acre can yield a hundred species of trees instead of a handful, and more species of insects than most of us can imagine. This understandable preoccupation with hot spots is a kind of biological version of Sutton's Law, named after the accomplished bank robber, Willie Sutton. When Sutton was once asked why he robbed banks, he allegedly gave the now classic reply, "Because that's where the money is." The biological "money" is in the hot spots, it seems.

My thoughts drifted back to an article published in *Science* in January 1997 by Princeton University's Andy Dobson and three coauthors, in which they analyzed the geographic distribution of endangered species in the United States.[1] Because their analysis of the data indicated that many endangered species of both plants and animals are found localized in a few hot spots of diversity—Hawaii, southern California and Arizona, southern Appalachia, and parts of the southeastern coastal region—they claimed that our endangered species can be protected most efficiently by spending conservation efforts and funds in these few key areas. I had been bothered by this article when it first appeared, and was still bothered by it as I looked at Stanley's and Donna's pictures. But then it was Joan's and my turn to show vacation photos.

Our album was open to a picture of a small, twin-engine Otter airplane disembarking thirteen warmly dressed people and waterproof bags, canvas Duluth packs, and partly disassembled canoes onto a flat expanse of gravel. Everything in the foreground lies in the gloomy shadow of early twilight, but a low, treeless hill in the background, yellow-brown in color, is brilliantly illuminated by the summer sun. Nothing in the picture is green. Not a single plant is in evidence. "It's very bleak," Judy remarked. "It looks like a desert."

It had felt bleak, I remembered, especially when the little plane, its tundra tires churning the lichen-covered stones, took off, climbed steeply out of the narrow Soper River Valley, and disappeared over the rounded hills of the Meta Incognita Plateau headed back to Iqaluit, once known as Frobisher Bay. We had been left at the upper reaches of the only canoeable river on Canada's vast Baffin Island, north of Hudson

Strait, in the eastern part of the newly designated Inuit territory of Nunavut. At eight in the evening, the Arctic summer sun was still fairly high in the sky but hidden behind a hill. A chill wind was blowing. Several members of our group glimpsed a wolf on the other side of the river. We found a partially sheltered spot off the gravel bar and pitched tents in the stunted vegetation while our Maine guides, Randy and Dwayne, cooked dinner on two Coleman stoves. That night we were awakened briefly by the sun shining on our tent at 3:00 A.M.

The next page of the album revives my spirits; it shows us two-thirds of the way up Mount Joy, with the Soper River winding through the valley below. "Everything is so brown," somebody comments. "Wait a bit," I respond, knowing what's coming in a few pages. Meanwhile, here is a shot of Mark showing off a hefty arctic char, kin to trout and salmon and the one species of fish in the freshwater part of the Soper or any other river on Baffin Island—a landmass that stretches farther than the distance from New York to Chicago. Char are carnivorous; the young eat the tiny shrimp and mosquito larvae that are among the few inverte- brates in the river, while the older char eat the younger ones as well. Not exactly an ichthyological hot spot, I muse as the page turns.

Next, Joan and I are shown running the first big set of rapids, which Randy has dubbed "Gatorade." What the Canadian geographer J. Dewey Soper called it we have no way of knowing. He doesn't give it a name in the first published description of the river named after him, which ap- peared in 1933. The page turns again, and suddenly everyone crowds around to get a better view. There are the thundering, ninety-five-foot falls of the Cascade River, a tributary of the Soper, with the giant block of ice miraculously clinging to the adjacent cliff. In the foreground is a patch of "mountain cranberry" dotted with last summer's tasty red berries. The only kinds of trees that grow on Baffin, shrubby willows and birches, most less than ten to fifteen inches high, cover the lower slopes of the valley of the Cascade. "How beautiful," Donna says, and I nod; when we were there I had told Joan that this was the most beautiful place I had ever seen. Soper, the first outsider to set foot in this valley with its tumultuous river and falls, called it "a fascinating commingling of savagery and soft- ness."[2] Another photo shows a dry upper hillside above the valley. Close- up photos reveal the mosses growing between the stones, with every rock sporting a multicolored array of encrusting lichens.

The page turns, and there we are hiking back from the falls to our third campsite. We see the tundra ponds surrounded by low, green vegetation, looking, if you squint a bit, like an aerial photo of a New Jersey salt marsh in June. I thought I saw tiny fish in one of those ponds, but when I brought Dwayne, a fisheries biologist, to look, he pronounced them shrimp. Then I got down close to them and noticed that they were swimming on their backs, just like the fairy shrimp that my teacher, Archie Carr, once took me to see in a temporary pond in north Florida.

Farther down the Soper River, at the fourth campsite, Joan and I stalked a caribou. This is easier than it sounds, in spite of the absence of cover; perhaps they don't see well, or perhaps they are unafraid and curious. The picture shows the caribou, looking too small for its ridiculously large antlers, browsing on plants in a gully about thirty feet from us. We could hear the crackling noises it made as it chewed. When we got back to camp, Randy, who tracks and studies bears for the state of Maine, pointed out a sandy patch where the ever-present caribou tracks were overlaid with the surprisingly big prints of a solitary wolf. The photos of the tracks came out well, but I was sorry that I hadn't listened to Joan and put my Swiss Army knife down next to them to give an indication of their size.

A few miles away, Randy directed us inland to the only two stands of real trees, willows, on Baffin Island. They grow in two adjacent patches of several acres each. Some of the willows are as tall as twelve feet. Dewey Soper found and photographed a similar patch, now dead, not far away in the valley. In this part of the Canadian Arctic, the tree line is four hundred miles to the south, in northern Quebec; Soper was very excited to come across actual trees that stood up from the ground rather than forming low, spreading mats. Joan thought that the location of the patches in hollows next to sheltering hillsides allows deep snow to accumulate, protecting the willows in winter from the freezing and drying Arctic winds. A photo shows Joan in the midst of the amazing forest. Only a few of the skinny trees are taller than her head. Another picture shows me beside one of two six-foot willows that Joan and I found next to a small stream in a steep-sided little valley several miles away. It is possible that no one else has ever seen these two trees. You can see that I am as proud as if I had discovered a giant sequoia anchored in the permafrost. (This is not as far-fetched as it seems. My friend John Tedrow, an expert on Arctic soils, has seen big logs, perhaps dawn redwood or larch, embedded in the Beaufort gravels of Prince Patrick Island's polar desert, far to the north and west of Baffin. They probably grew there several million years ago when the climate was warm.)

More exclamations from the people around the table. We have come to the pictures of arctic flowers blooming in what Soper described as "unrivalled loveliness . . . in the briefest, though the fairest summer season in the world."[3] Joan has marked each plant in the album with its common and scientific name. Here is a lowland wet spot covered with eight-inch stalks of the cottongrass sedge, *Eriophorum angustifolium*, its cottony tufts of plumed seeds waving in the breeze. Not in the photo are the billions of seeds drifting by in the sky above the ridgetops, making the light refract and sparkle as if the air were filled with tiny diamonds. On the next page are the ubiquitous, showy purple flowers of arctic willow herb, *Chamerion (Epilobium) latifolium*, and underneath on the page, a rounded cluster of yellow mountain saxifrage, *Saxifraga aizoides*, looking like an arrangement in a florist's shop. Then comes a picture of mountain avens, *Dryas integrifolia*, with Joan's Swiss Army knife looming over it; and here is a clump of white heather, *Cassiope tetragona*, that has nearly finished flowering in late July—a few white blossoms can still be seen among the seed capsules.

The major study of Baffin vegetation identifies only 189 species of higher plants on the entire island, an area larger than the British Isles. In our seven days on the river, Joan found and keyed out 54 species, a significant part of the flora. She usually spent an hour or so before dinner looking through her hand lens at each of the day's plant specimens spread out in her lap, while others in the group explored, rested, or gathered around Tom while he sang and played the guitar.

The album ends with pictures of some of the beautiful and vivacious Inuit children who accompanied us everywhere on the last day of our trip while we were in Kimmirut, formerly Lake Harbour, on the southern coast. There is also a picture of three skulls of polar bears—killed in self-defense—resting on top of a shed; and a final view of the Meta Incognita peninsula, brown, lifeless-looking, and laced with meltwater streams and narrow patches of snow, as we flew over it on our way back to Iqaluit.

Joan closed the book. I could see that the people around the table were moved by what they had seen, but there was no way that Joan and I could convey the depth of our feelings about this place, at once so vast and unforgiving, yet so intimate and welcoming; so poor in species diversity, yet so vibrant with life. What we encountered in Baffin Island, and what made it magical, was that increasingly rare quality of nature not yet vanquished, an amazing and nearly pristine ecosystem still in place, still working. It is not a hot spot, it is an evolutionary pinnacle of a different sort, nature fully equal to the terrible rigors of a harsh climate.

Why, I wondered, has species diversity become a god to conservationists? The wealth of species that adorns the tropics is one of the greatest glories of this planet; it must be protected. But wealth of species is only one way of reading nature; there are many others. A preoccupation with hot spots can absorb too much of our energy and concern, leaving little for the rest of the natural world. Unspoiled places like the valley of the Soper River, poor in species as they are, have their own kind of grandeur and glory, a living presence that surely makes an equally powerful claim on our care and affection.

Nevertheless, the idea of hot spots persists and has now captured the public's attention. The eminent British conservationist Norman Myers, who first used the term back in 1988, published a paper about it, with four coauthors, in the February 24, 2000, issue of *Nature*. The twenty-five principal hot spots in the world, mostly tropical, contain "as many as 44% of all species of vascular plants and 35% of all species in four vertebrate groups" in an area that amounts to "only 1.4% of the land surface of the Earth," Myers claimed. The word "only" is an important one; it leads to the key idea in the paper, which in turn is probably the reason why *Nature* gave such prominence to this article about hot spots.[4]

The idea is simple and is based on a fact that is hard to impeach: there are limited funds available for conservation in the world. Given this fact, say Myers and his colleagues, "How can we protect the most species per dollar invested?" The answer is simple, we need to implement "a 'silver bullet' strategy [a specific remedy precisely targeted to a particular problem] on the part of conservation planners, focusing on these hotspots in proportion to their share of the world's species at risk." Stuart Pimm, an ecologist at Columbia University who coauthored *Nature's* commentary on the Myers paper, was also quoted in a March 14,

2000, *New York Times* article on hot spots written by William K. Stevens. "They're probably going to capture pretty much 30 to 50 percent of everything," said Pimm, referring to a silver bullet campaign to save hot spots. The Myers study was done, he said, with "extraordinary care and numeric detail"; the paper is "one of the most remarkably optimistic statements about biodiversity that we have ever seen." Russell Mittermeier, the president of Conservation International, an expert on primate conservation and second author of the Myers paper, was also mentioned in the *Times* article. Mittermeier referred to their statement in the paper that $500 million spent annually would "go far towards safeguarding the hotspots." This is an amount of money that is only twice the cost of a single Pathfinder mission to Mars, and vastly less than the $300 billion annual cost of protecting worldwide biodiversity that had been previously estimated. Five hundred million dollars, Mittermeier observed to the *Times*, is an achievable goal.

So for $500 million per year, not even the cost of one Stealth bomber, we can begin to protect as much as one-third to two-fifths of the earth's biodiversity. There are many individuals who could part with $500 million without any appreciable change in lifestyle; there are multinational corporations for which $500 million is less than their advertising budget; there are governments for which $500 million is an insignificant fraction of their annual expenditures.

It is no wonder that for Myers, Mittermeier, and Pimm this news is cause for great optimism, and their hopeful message has been conveyed by the newspapers and public radio to many hundreds of thousands of people. Theirs is a feel-good message because it means that countless rare species might be saved. Not only will these complex ecosystems continue to function, but somewhere inside them may well be living the species of plant, microorganism, or even animal that will end the scourge of cancer, provide the next generation of antibiotics, give us new miracle pesticides, inspire more inventions such as Velcro, and maybe—if we're lucky—cure baldness.

But there is another, more important, reason why the message of hot spots makes us feel good; it reassures us that we, in the richest and most powerful of countries, don't have to make any fundamental changes in the way we live our daily lives, the way we behave toward poorer nations, and the way we treat the other 98.6 percent of the land (not to mention the two-thirds of the planet that is ocean) in order to save biodiversity. With a trivial expenditure to safeguard the hot spots, an expenditure that is much less than 1 percent of the annual gross global product, we can keep on with business as usual.

Toward the end of the Myers article, the authors discuss conservation funding. They claim that during the ten years between 1989 and 1999, nongovernmental organizations such as the MacArthur Foundation, Conservation International, and the World Wildlife Fund spent an average of $40 million per year on hot spots conservation. This, they assert, "is only a tiny fraction of the amount spent per year on biodiversity conservation by governments and international agencies, these funds being assigned mainly to across-the-board activities rather than the concentrated efforts advocated here." Although they describe their "silver bullet" strategy of concentrating on hot spots as complementary to other conservation activities, their description of these other efforts as "scattergun . . . seeking to be many things to many threatened species," implies that the hot spots strategy is a better alternative, not a complement. We must set conservation "priorities," they claim. "How can we support the most species at the least cost?" they ask.

I take it as established fact that there is now a mass extinction taking place as a result of human activities; that the bulk of the species extinctions are taking place in the tropics, because that is where most of the species are; that there are limited funds available for international conservation programs; that to the extent that these programs work, a conservation dollar spent in a hot spot is likely to save more species than one spent elsewhere; and that if we don't save biodiversity now, it will not regenerate during the future course of human history. Nevertheless, I am unhappy with the hot spots concept—both with the way it was presented by Myers and his colleagues and with its subsequent popular development.

The main problem with hot spot–style conservation is shown by a look at the map of hot spots presented by Myers, which reveals a striking feature. Most of the terrestrial world is not hot spots; there are many large countries, such as Argentina, Mongolia, Canada, Zaire, and Poland, that have no hot spots in them. If the Myers team was simply pointing out places that are especially rich in biodiversity and are deserving candidates for conservation, this wouldn't matter, but it wrote of limited conservation resources and the need for setting priorities. In other words, if we have to choose between the Philippines and Canada, the Philippines wins. Or Brazil's Atlantic forest versus Argentina, with Brazil winning. The tropics versus the temperate and boreal regions, with the tropics winning. Don't we have enough problems without making conservation a nationalistic battleground? Even within countries, this is a divisive strategy. The California floristic province is a wonderful place, but would we really want to divert conservation resources from the New York–New Jersey Highlands, the Flint Hills of Kansas, or the Maine woods to this western hot spot?

A less obvious but equally important flaw in the hot spots approach is that it treats the value of nature as a numbers game. A density of 5,000 species per square mile is better than 2,500; 10,000 is better than 5,000. I suppose that this comes from our number-obsessed, science-and-technology-obsessed culture, in which anything that can't be quantified is suspect. But there are other ways of looking at the things that matter, as the trip to Baffin demonstrated. There's more to nature than the number of species in it. Moreover, although the tree diversity in the tropical hot spots is much higher than it is in my native New Jersey, for example, few of those tropical trees would survive a New Jersey winter. Wonderful as the hot spots are, the rest of the world is not going to be biotically maintained by their conservation.

Before we devise global conservation solutions—if that is what we are trying to do—we ought to step back and ask what is causing the biotic impoverishment that these solutions are attempting to address. I don't mean to imply that there is only one cause; there are many, and they are intertwined. But there is one that stands out for me, globalization, which is the divorcing of business from any association with place or community. Globalization as a cause of the problem is getting obscured by globalization as a purported cure, a situation that must be avoided at all costs.

To explain what I mean, I want to look at some of the implications of the hot spots strategy. The approach I am going to take was first suggested to me by one of my graduate students, Kristi MacDonald-Beyers. I was discussing the Myers paper with her, and she remarked that hot spots are the conservation equivalent of globalization. I had thought a good deal about globalization, although never in that context, and I immediately got a sense of what she meant. Then, by chance, a few days later, I came across a featured column in the *New York Times* that made it perfectly clear. It was written by Thomas Friedman, a strong supporter of globalization, on March 31, 2000. Friedman criticized the then-forthcoming protests in Washington against the International Monetary Fund and the World Bank. "[T]he truth is the I.M.F. and World Bank . . . are bit players, of declining importance, in a much larger drama. . . . The real drama is on the ground in the developing world . . . [and] the real players are a diverse collection of companies, indigenous peoples, endangered species and local governments" who, he reports, are working together to save the Guyana Shield, the world's largest unbroken expanse of tropical rain forest. Friedman was there with a team from Conservation International, which was brokering a deal to put together a 13-million-acre protected area in the Venezuelan part of the Shield. The deal involved debt-for-nature swaps and the participation of Venezuela's national oil company, Petróleos de Venezuela, which owns Citgo in the United States. Why was Petróleos participating? Because, according to Friedman, they wanted to enhance their global environmental image. Friedman quoted Glenn Prickett, senior vice president of Conservation International, as saying: " 'Thanks to globalization and the Internet, power is now much more diffused, global companies are now much more exposed, and organizations like ours much better positioned to offer solutions.' " So don't just protest, Friedman counseled the Washington protestors, go out and do something positive to help globalization save the environment.[5]

I am glad that Conservation International is working in Venezuela to protect the Guyana Shield and the Yek'wana and other indigenous peoples who live there. Fortunately, their efforts go far beyond the agreement with Petróleos.[6] However I am less happy with the projection of this kind of deal as a model of effective, globalized conservation. Debt-for-nature swaps have not been overwhelmingly successful; they depend on too many uncontrollable factors, including the value of the national currency, which has to remain low without becoming absolutely worthless for a long period of time if the swap is to succeed. Moreover, political factors concerning land ownership can complicate and damage debt-for-nature swaps. I also wonder whether Petróleos, which has enlisted the World Trade Organization to undermine U.S. environmental laws, will care so much about its environmental image if its technical advisers should discover oil, diamonds, gold, silver, or uranium in the reserve; or what will happen if the government decides—as governments do—to scrap the deal. But these are not my major concerns.

The connection of hot spots with globalization, as Friedman should realize, is that for his readers, *the hot spots are almost always somewhere else.* That may not bother him, but it bothers me. When I was in Baffin Island, I understood the tremendous importance of conserving its magnificent ecosystems, but I never thought that they should be saved instead of the ecosystems of my home, New Jersey. After his brief visit to the Guyana Shield, Friedman came back to New York State, whose endangered species and threatened habitats are not displayed on his radar screen. That's the nice thing about hot spots for many supporters of globalization: they don't have to live in them day in and day out. (The same can be said about carbon offsets—our $CO_2$ pollution is supposedly mitigated by the growth of forests elsewhere, but the elsewhere is often comfortably out of sight.)

Globalization, however it is rationalized, is not the free and equal trade of goods among prosperous nations everywhere, nor will it be. It is often the use by multinational corporations of cheap, semi-slave labor in poor, mostly tropical countries to grow cash crops and make goods that are sold in the rich, temperate zone countries—at least for the few more years that those countries remain rich. The economist Herman Daly and the theologian John Cobb have written in their book, *For the Common Good,* "Free traders, having freed themselves from the restraints of community at the national level and having moved into the cosmopolitan world, *which is not a community,* have effectively freed themselves of all community obligations."[7] Many of the consequences of this uncoupling have been terrible, and a number of us feel guilt over the global activities of some of the largest corporations—guilt which is enhanced by our enjoyment of our transient wealth. Enter the hot spots. We can now save nature while driving our SUVs. But are we saving nature (even hot spots) with this hands-off, uninvolved, globalized conservation? A group of twenty-one highly regarded conservation biologists from the United States, Central and South America, and Asia, writing in the journal *Science,* have detailed the objections to the globalization of conservation:

Generalized global approaches fail for biodiversity conservation at local scales, because solutions must integrate extremely diverse natural, socioeconomic, and cultural systems and usually require a sense of community ownership. . . . We argue that leadership in conservation has to be decentralized and better integrated into local conditions. Locally produced strategies and agendas, implemented by strong local institutions and individuals are key to success.[8]

There is a strong disconnect between "locally produced strategies and agendas" and "strong local institutions" on the one hand, and globalization—one of the inheritors of colonialism—on the other. Although it is not the intent of the conservation globalizers, their approach, which is largely "western" and "northern" in origin, can be condescending toward the local cultures that, after all, preserved the hot spots until the onset of globalization. Here is an extract from Barbara Kingsolver's extraordinary historical novel, *The Poisonwood Bible*; it makes the point well, albeit for an earlier age.

[The first Portuguese explorers] reported that the Africans lived like kings, even wearing the fabrics of royalty: velvet, damask, and brocade. Their report was only off by a hair; the Kongo people made remarkable textiles by beating the fibrous bark of certain trees, or weaving thread from the raffia palm. From mahogany and ebony they made sculptures and furnished their homes. They smelted and forged iron ore into weapons, plowshares, flutes, and delicate jewelry. The Portuguese marvelled at how efficiently the Kingdom of Kongo collected taxes and assembled its court and ministries. There was no written language, but an oral tradition so ardent that when the Catholic fathers fixed letters to the words of Kikongo, its poetry and stories poured into print with the force of a flood. . . . The Europeans were dismayed to find no commodity agriculture here. All food was consumed very near to where it was grown. . . . The Kingdom was held together by thousands of miles of footpaths crossing the forest, with suspension bridges of woven vines swinging gently over the rivers.[9]

And biodiversity thrived. I am not suggesting that the Congo go back to the way it was before the Portuguese came. That would be impossible. Nor do I advocate abandoning conservation efforts in the tropics. Nevertheless, it is time to admit that conservation is much more than a matter of throwing dollars and experts at the primarily poor, primarily tropical places that have the most species. Maybe asking which areas of the world should receive the highest priority for preservation is not a good question. A better one is, "How can we live in a way that reduces the burden on species and ecosystems elsewhere?" Another good question is, "How can we live so that we protect the species and ecosystems in our own community?" Perhaps these are the same question; certainly they have the same answer.

The plant geneticist and ecologist Wes Jackson, cofounder of the Land Institute in Salina, Kansas, has increasingly turned his attention to the question of how to live successfully in a community. In his book, *Becoming Native to This Place*, he writes, "An extractive economic system to a large degree is a derivative of our perceptions and values. But it also controls our behavior. We have to loosen its hard grip on us, finger by finger. I am hopeful that a new economic system can emerge from the homecomer's effort—as a derivative of right livelihood rather than of purposeful design."[10] More than two decades ago, Wendell Berry had the same message in *The Unsettling of America*: "The possibility of the world's health will have to be defined in the characters of persons as clearly and as urgently as the possibility of personal 'success' is now so defined. Organizations may promote this sort of forbearance and care, but they cannot provide it."[11]

How do we make personal forbearance and care a force in the world? Wes Jackson says that one of the most effective ways for this to come about "would be for our universities to assume the awesome responsibility to both validate and educate those who want to be homecomers—not necessarily to go home but to go somewhere and dig in and begin the long search and experiment to become native." Berry has more immediate suggestions, published in the *Amicus Journal* in 1992. A few of his recommendations are to buy near at hand, rather than at a distance; buy from a local store rather than a chain store; do not buy anything you do not need; do as much as you can for yourself; if you can't do it for yourself, see if a neighbor can do it for you; buy food that is locally grown; grow food for yourself; take your vacations as close to home as possible.[12]

The relationship between conservation and home has been explicitly and eloquently spelled out by the Pennsylvania State University ecologist Chris Uhl:

> The ultimate challenge for conservation biology isn't to create more parks, manage more forests, or shepherd through more environmental legislation. Rather, the challenge is to change the way we as a society perceive the natural world. Einstein was right: "We shall require a substantially new manner of thinking if mankind is to survive." The parks and enlightened legislation will come rushing in once we get our thinking right. . . . As mundane as it may sound, for many of us the land at our doorstep provides the starting point for developing an affection for the earth, which is a necessary foundation for living respectfully within the confines of our planet. The healing of the earth will spread outward from the land around our doorsteps.[13]

There is something very special about local conservation, something that operates at the most profound level of human existence, something that is lacking when we think about global conservation, important as it is. Whenever I drive along the Millstone River, here in central New Jersey, I am reminded of a particular night in January 1976. Earlier that day, my nine-months'-pregnant wife, Joan, who hates sitting around, came home early from work and shoveled out most of our one-hundred-foot driveway after a snowstorm. Now, at 3:00 A.M., we were on our way to the Princeton Medical Center for her to give birth to our second child, Jane. The snow had stopped, and the stars were out. Our road ran for miles along the Millstone River, which was peacefully flowing under the ice past snowy fields and woods whose pin oaks, river birches, and box elders stood out stark and beautiful against the white background. It may not be a hot spot, but the memory of the river and trees on that winter night is an indelible part of me. Whenever I see them again I feel a deep sense of gratitude that they are still there, thanks to the efforts of my state's many conservationists. Any threat to them would be a threat to my own person.

The force of such memories and associations cannot be exaggerated: the weedy, urban lot next door where at the age of five you first saw a monarch caterpillar on a milkweed leaf; the mating calls of peepers and spadefoot toads coming through your open bedroom window on a spring evening; the smell of the pine woods where you may see a richly colored corn snake sunning itself on the sand road while you walk your dog. Most of us have memories like these, even if we don't know it. Local conservation has continuity; it is passed on from parent to child, friend to friend, and it has a strength that has to be seen to be believed.

We must not lose our appreciation of hot spots, but we can extend the concept to rich natural habitats everywhere. One such habitat was described to me by my friend Frank Lake, an American Indian and an experienced field biologist from northern California. Frank has told me that in the Indian country that he knows best, the sites that the native peoples held most sacred, and which they managed and protected with the greatest care, appear to be the places that had the highest number of species and the highest natural productivity. In Frank's words:

> My father and I would visit the high country in the summer months between the time when I was eight and thirteen years old. It was at this time I learned the importance of being a person of place, and what it meant to have healthy and productive watersheds. Our visit was almost always for religious purposes. I was taught that everything in this area was sacred and that it must be respected. Respected even if I did not understand it, and that in time, if I observed what was going on around me, and I was patient, I would eventually learn to understand and value the power of place.[14]

If we think of hot spots as relative, not absolute, then every place has its hot spots. This home-centered application of Myers's important concept is good biology, good conservation, and good community service. The sacred hot spots identified by Frank Lake's father were conserved by *local* people to meet *local* spiritual and physical needs, the two of which were inseparable. One cannot ask for better conservation or a more durable economics than that.

---

## PUTTING A VALUE ON NATURE

I had a call from Bill Stevens at the Science Desk of the *New York Times*. "I'm doing a story about an article that is about to appear in *Nature*," he said, "and I'd like to fax you a copy for your comments. The article is by Robert Costanza and twelve other economists, geographers, and ecologists.[1] It's called 'The Value of the World's Ecosystem Services and Natural Capital.'" As soon as I heard the title, I knew that I was not likely to see eye to eye with the authors. With some reluctance, partly because Costanza is a committed and accomplished environmental economist whom I respect enormously, I agreed to read and comment on the article. Soon I had it in hand.

I started out to read the article thoroughly, word by word, examining the methods, the assumptions, and the numbers in the tables with care. But it was hard to do; I kept wanting to skim. Not that it was badly written; it was clear and lucid. Not that it was aggressive or arrogant in tone; indeed, it was mildly apologetic. Not that it was ill-intentioned; its goal was to demonstrate that nature in the form of intact ecosystems is worth far more to us than we currently imagine. Nor were the estimates exaggerated; as the authors claimed, they were obviously understated, giving a minimal figure of $33 trillion for the annual value of nature. This is approximately twice the value of the global gross national product, they pointed out. In other words, nature is very valuable to us, even more valuable than the goods and services that people produce.

One problem was immediately obvious. The authors did not seem to realize the dangers of what they were doing. Never mind the larger dangers to our essential humanity, but even the practical, economic dangers. Suppose, for example, we are able to increase greatly our global gross national product, perhaps to $50 trillion or $75 trillion?

Should we care less for nature then? Or, what if a group of antienvironmental economists calculates the global *costs* of nature's harmful activities: the floods, the droughts, the damaging winds, the insect pests, the weeds, the infectious diseases and their vectors? What if these calculated costs amount to $34 trillion?

But something else was wrong, more fundamentally wrong. I looked again at the text. Nature was reduced in this analysis to seventeen "ecosystem services and functions," including "climate regulation," "water supply," "soil formation," "waste treatment," "pollination," "food production," "recreation," and "cultural." Adding the estimated values of these seventeen items gave Costanza and his coauthors their figure of $33 trillion dollars, enabling them to compare the worth of nature's services with the global gross national product. The reductionist absurdity of this Swiftian exercise first irritated, then amused me. What really got to me, however, was the feeling that the effort to assign economic values to particular functions of nature was somehow terribly misguided in a different way, even if I couldn't put my finger right on the problem.

I faxed my comments back to Bill Stevens, and a few of them were duly printed in his story. "Common sense and what little we have left of the wisdom of our ancestors tells us that if we ruin the earth we will suffer grievously," he quoted me as saying, followed by the statement, "I am afraid that I don't see much hope for a civilization so stupid that it demands a quantitative estimate of the value of its own umbilical cord."[2] Stevens slightly softened the bluntness of my remarks by pointing out that I accepted Costanza's numbers as conservative estimates. And that, I thought, was that. But it wasn't, really—this kind of argument stays with you, stewing quietly as you go about other business. The weekend came.

On Sunday, Joan, Sam, and I bought a gingko tree. There was a large gap, right in front of our house, in the line of trees growing along the street. The city had offered to plant a tree in the empty space, but I didn't want any of their white ashes or Japanese *Zelkovas* shading out the incredibly productive sour cherry in the lawn next to my house, so I said no. I wanted a gingko, with its beautiful leaves, its wonderful history as a living Mesozoic fossil rediscovered in a Chinese temple garden, its high tolerance of street pollution, and its growth form compatible with the neighboring cherry tree. After trying four different nurseries, we found one with a perfect gingko: only five feet tall so that Joan and I could move it ourselves, and with a natural gingko shape, unlike the skinny, columnar varieties that are commonly available. A grafted male, it would not be dropping any vomit-smelling fruits on the sidewalk in the fall. The nurseryman said that it was a variety known as October Gold. It cost $89.95.

On the drive home, I thought about the value of the piece of nature that we had purchased. What was it worth to us now? Clearly $89.95 because that is what we had just paid. Would we have spent $179.90? No, that would have been too expensive. Maybe, I thought, Costanza is right. We can assign values to nature; we do it all the time. But people have grown accustomed to taking nature's services for granted (like the services of housewives). Why not wake them up by showing how much these services are really worth? All right, I answered myself, show them what nature is worth, but why do it in dollars? That's the enemy's turf. There the internal argument rested, undecided.

In twenty minutes we arrived home and carefully wrestled the gingko, with its heavy, burlap-wrapped root-ball, out of the station wagon to the spot where we intended to plant it. The sun illuminated its little May leaves and reflected off the plastic $89.95 price tag. I wondered whether I should have sent those comments to Bill Stevens.

The time was 11:15. I had to drive Sam to his soccer game at 2:30. No problem. I had more than three hours to dig a hole about twice the size of the root ball, put in the gingko, and fill the spaces with dirt from the hole mixed with a couple of bags of high-quality topsoil. Three-quarters of an hour would be ample.

The spot I had chosen was a slightly depressed, circular area about two feet wide where the grass had never grown properly. On the first swing of the pick, I hit something solid that gave off a chunky sound. A few more strokes of the pick and some shoveling uncovered the problem. It was a large buried stump, probably from a silver maple. The stump had been there for at least the fifteen years we had owned the house. "Move the hole a few feet away," Joan suggested, but that sound advice convinced me that no other spot was suitable. Besides, I reasoned, if the stump has been in the ground so long, it was probably rotten and would be easy to remove. "Have it your own way," said Joan, and she went off to plant some flowers along the front walkway.

I started in with the pick and a solid, straight-bladed, Irish garden spade, using both to dig around the stump and to break off pieces of it. The trench around the stump deepened to only three or four inches before I hit roots and other obstructions. The tiny pieces of stump that I was able to break loose would have embarrassed an ailing chipmunk. Time passed and I labored on, getting first hot, then exhausted. Sam and Joan took pity on me and joined in. Sam was delighted to get a chance to use the pick.

More time passed; we all grew tired. We could see four or five inches of the stump in some places, but the bottom of it remained well hidden deep in the earth. Joan went back to her flowers, and Sam and I continued, taking turns with the pick. I was now angry at the stump and at myself; most of my pick strokes were missing the mark. Then Sam said, "Dad! You're not doing anything. Why don't we try to split the stump?" Grateful for the idea, I leaned on the spade while he went off to get a sledgehammer and wedge. We worked at this for a little while and made a little more progress. Occasionally a two-inch piece of stump would break off, but the wedge got stuck frequently and had to be loosened with tiring, sideways blows of the sledge along with strenuous pick work. My watch said 1:00. At the rate we were going, it would take another four hours or more. Sam reminded me that the soccer game was still at 2:30.

Then my neighbor, David Freedman, came by. He is a practical man who was raised on an anarchist agricultural commune and has owned and run factories for the majority of his more than seventy years. "Burn it," he advised, handing me a bottle of waste gas-oil mixture and some matches. Soon the stump was blazing merrily, while I looked around nervously for the police. Every now and then I would squirt some charcoal lighter on it, and I even added some charcoal to increase the heat. By 1:45, the stump seemed only lightly singed, Sam was openly skeptical, and I remembered that the gas line from the street to our house ran only ten feet away. Lugging over the garden hose, I put the fire out. Not quite a quarter-inch of the wretched stump had been consumed. The remainder seemed fire-hardened. Wearily, I picked up the sledge while Sam positioned the wedge.

A shadow fell across the stump. I looked up. It was my new neighbor from across the street. I had forgotten his name (Scott). He was holding a black iron bar, five feet tall, with a blunt, inch-and-a-quarter-wide chisel blade at the bottom. The bar tapered to a sawed-off circular striking surface about five-eighths of an inch wide at the top. "Try this," he said.

"What is it?" I asked.

"I think they call it a shale bar. I found it in the garage when we moved in." It was very heavy, about twenty pounds. I came down hard with it on to the stump and drove it further in with a few blows of the sledge. The stump split, and the length and weight of the bar let me easily pry the two halves apart. By 2:00, the stump was lying on the sidewalk in four massive pieces. Twenty minutes later, the gingko was planted and Sam and I were off to the soccer game.

The next day, I went to our local masonry and hardware dealer to buy a shale bar of my own. There were a half-dozen of them standing in a corner along with a different type of bar that was a little taller and thinner and had a curiously expanded, pancake-like top. "I would like one of these shale bars," I said casually to the man behind the register.

"That's a pinch bar you've got there. The shale bars are the other ones."

"What's the difference?"

"The pinch bar is heavier. Believe it or not, that makes it easier to use. Our customers mostly break up concrete with them. The shale bars are for going through shale—like if you were excavating for a swimming pool. And the pinch bars are five dollars more. That'll be $34.99, plus tax."

At home, I looked at the pinch bar more closely. Sighting along the length of it, I noticed visible irregularities that made it look handmade. The label said "Rat Tail Crow Bar"—an interesting zoological hybrid—"Cotter & Co., Chicago, IL, Made in USA." On a whim, I called Cotter & Co.

"We're a distributor, not a manufacturer," said the second person I spoke with. "That bar came from the Mann Edge Tool Company, in Lewistown, Pennsylvania." I thanked her and called Mann Edge.

The man I talked with in Lewistown was able to tell me a bit more. "Those are high-carbon steel bars," he noted, "made in a rolling press. But we don't make them. We get them from the J. & H. Company, in Columbiana, Ohio. You should call them and talk with John Kephart."

When I finally got through to John Kephart, I learned that he was the "J" of J. & H., the "H" being his wife, Helen. "We're one of the few plants left in America that make those bars," he told me. "We've survived because this is a small, tightly run family operation with a low overhead and no retail sales expenses. Our furnaces are gas-fired, and we recycle our solutions, so we don't have a pollution problem. Your bar was made from 1060 high-carbon steel in an open die, using a piece of equipment known as a forging roll. Our presses were mostly made in the teens on into the 1940s. We have to be able to cast parts and fix them ourselves."

As I had guessed, forging with an open die involves much hand craftsmanship and operator skill. The pinch bar starts out square in cross section near the bottom and gradually becomes round farther up. That's hard to achieve. It takes about two years to learn how to make a bar.

"We're proud of their quality," Kephart said. "Also, we refuse to temper the ends of them, as some companies do. That would make the bar harder, but might introduce microscopic heat cracks, causing the bar to break under stress. So our bar will only last you ninety-five years, five years less than a tempered bar."

I learned from Mr. Kephart that pinch bars are mostly used for positioning railroad ties and moving heavy machinery. From his words, it was clear that their use, like their manufacture, is an art. I enjoyed talking with him—here was an expert, I thought, whose knowledge was useful and trustworthy.

After the call, I stepped outside to admire the gingko tree and removed its $89.95 price tag. Then I went over to survey the pieces of the stump piled by the driveway. What was the value of the stump, I suddenly wondered? Not a great deal as firewood. Was there anything else? Well, it had given Sam and me a chance to work hard together—and it had made me listen to his advice. The physical exercise was surely beneficial. I had strengthened my ties to two excellent neighbors while involving them in a task that improved the neighborhood—restoring a tree to a place where one had been before. I had gained understanding of a tool and a craft that belonged to America's history, and, I hope, to its future. Yes, the stump was worth a great deal more than the purchase price of the gingko, but the actual amount would be impossible to calculate. Even nature that has been fought, dismembered, and discarded has its value. So much for attaching a dollar price to nature, I thought, as I carried my new pinch bar to its place in the garage.

---

## THE DOWNSIDE OF CORPORATE IMMORTALITY

The example of the American States shows that private corporations—whose rule of action is the interest of the association, not the conscience of the individual—though composed of ultra-democratic elements, may become the most dangerous enemies to rational liberty, to the moral interests of the commonwealth, to the purity of legislation and of judicial action, and to the sacredness of private rights.

George P. Marsh, *Man and Nature*[1]

On May 22, 2000, Randolph Walker, a seventy-one-year-old career actor, was struck and killed by a double-decker tour bus while crossing a street in midtown Manhattan near the low-rent apartment complex where he lived. This sad accident—an all-too-common occurrence in crowded cities—nevertheless galvanized the New York community, for reasons that should concern us all.

Mr. Walker was on his way home after an audition for a bit part in a movie called *Dummy*. He had been hopeful about getting the part; he was always hopeful. This is probably why he had been able to eke out a living as an actor for nearly twenty-five years after leaving a job in banking. He pieced together bit parts in plays, work in dinner theaters, and "under five" roles in soap operas. "Under five" roles are those in which an actor is hired for one day and has less than five lines to speak. In his obituary in the *New York Times* (June 1, 2000), reporter Robin Pogrebin quotes a friend, Todd Heughens, as saying: " 'There are so many actors who don't work at all, and here was a guy who supported himself as an actor. . . . So I think he took great pride in that.'" Yet he was described by his fellow actors as "non-competitive in the extreme. . . . He told his friends about roles he was seeking so they could try to get them also," and "he was happy for them even when they won parts he had wanted."

Mr. Walker, who was six feet tall and slender and had a resonant voice, was generally cast in small character parts, such as butlers or undertakers. He loved acting and was an inspiration and mentor to younger actors, according to his agent, Michael Hartig, who was also quoted in the obituary: "'When he worked on shows, they would gravitate toward him,' Mr. Hartig said. . . . 'He always had time to talk to them, to share his knowledge and his expertise.'" He was a quiet, unassuming professional who lived for his work, did not resent his lack of fame, and left behind nothing but good reports and sadness when he died.

Ironically, it was the manner of his death that secured Mr. Walker a firestorm of public sympathy and a major obituary in the *Times*. The driver of the bus that struck him was unlicensed to drive it, and Mr. Walker had the right of way. Later, an investigation by city and state agencies of the bus company, New York Apple Tours, found a shocking history of hundreds of violations of city, state, and federal laws and regulations, and it set off a wave of outrage against the company.

Nine days after Mr. Walker was killed, Governor George Pataki ordered the State Department of Motor Vehicles to suspend the vehicle registrations of all sixty-three of New York Apple's buses. "We can't tolerate an operation that ignores the regulations and flouts the law," the governor said. "They should be, in my view, put out of business." Mayor Rudolph Giuliani agreed. But on June 14, a state Supreme Court judge, Phyllis Gangel-Jacob, reversed the governor's order on the grounds that it would irreparably harm Apple Tours—and the company was back in business. The city and state appealed the decision, and on December 14 a five-judge panel of the Appellate Division of the State Supreme Court reversed Judge Gangel-Jacob, temporarily banning Apple's buses from the streets again. Apple Tours appealed this judgment to the state's highest court, the Court of Appeals, and in January 2001 that court dismissed the suit, allowing the suspension to stand. On February 8, Apple Tours surrendered its operating license and, according to its lawyer, went out of business. By that time, it had been fined more than $8 million and was selling off its assets, including about a dozen double-decker buses purchased by a division of Coach USA.

Despite these events, State Senator Thomas K. Duane, who worked with Manhattan community groups to try to shut down the company, was unwilling to declare victory. "They always have another angle," he told Sherri Day, a reporter for the *New York Times*, who published her story on February 26. "You can't trust what they say, and you can't trust what they do. We still want to have a permanent revocation of their license, because no matter what they say, we want to make sure that they'll never be able to do business in the City of New York again." Presumably, Senator Duane was gratified when the New York State Department of Transportation permanently revoked the bus company's charter in April. As reported by Ms. Day in the *Times* on April 9, a spokesman for the state agency said, "If the principals wanted to reestablish Apple Tours or another company, they would come under very close scrutiny."

A Republican governor and a Republican mayor helping to put a corporation to death for its crimes! Don't they know that for more than a century, corporations—at least the very big ones—have been immortal, shielded by our system from threat of termination? Their names may change, their officers come and go, they may merge or split, reorganize, or go bankrupt, but the corporation lives on, its "rights" intact, regardless of what it does, forever.

It wasn't always this way. Until the last third of the nineteenth century, corporate charters could be and were revoked, for cause, by the states that granted them. Even as late as 1890, Justice Finch wrote for a unanimous N.Y. Court of Appeals in the case of the North River Sugar Refining Company, "The judgment sought against the defendant is one of corporate death. . . . The life of a corporation is, indeed, less than that of the humblest citizen." Where corporate transgression "threatens the welfare of the people," Judge Finch continued, "they may summon the offender to answer for the abuse of its franchise or the violation of its corporate duty." The judgment of the court was to order the immediate dissolution of the company.[2]

The U.S. Constitution doesn't mention corporations. The idea of corporate immortality in U.S. law dates back to 1819, with the decision of U.S. Chief Justice Marshall in the case of Trustees of Dartmouth

College vs. Woodward. Marshall spoke of "a perpetual succession of individuals . . . capable of acting . . . like one immortal being." But as the chief justice also said in that decision, a corporation is "a mere creature of law, it possesses only those properties which the charter of its creation confers upon it." Presumably, if a corporation violates its charter, the law can unmake it. Its immortality is conditional.[3]

All that has changed. For the largest corporations, charters have become little more than legal formalities, and corporations have rights and powers in court greater than those of any individual citizen. People all over the United States have learned this to their sorrow, for example, while fighting in court to stop the encroachments of cellular towers and supermalls on landscapes in their communities. As George Marsh wrote more than a century ago,

> More than one American State is literally governed by unprincipled corporations, which not only defy the legislative power, but have, too often, corrupted even the administration of justice. The tremendous power of these associations is . . . [due partly] to an old legal superstition—fostered by the decision of the Supreme Court . . . in the famous Dartmouth College case—in regard to the sacredness of corporate prerogatives. There is no good reason why private rights derived from God and the very constitution of society should be less respected than privileges granted by legislatures.[4]

With the emergence of the World Trade Organization, the little chance that courts and legislatures will control the largest corporations has all but disappeared. A three-member panel of the WTO, meeting in secret, invalidated a section of the U.S. Clean Air Act on behalf of the Venezuelan oil industry. The WTO has also ruled against the part of the U.S. Endangered Species Act that protects sea turtles from needless death in the nets of foreign shrimpers. And it has sanctioned the European Union for its ban on the hormone-laced beef produced by U.S. agribusiness. Thanks in part to the unrelievedly pro-multinational rulings of the WTO, it is our laws and regulations, and those of the European Union, that are being weakened, not corporations.[5]

Did the governor realize what he might be starting when he sought the death of New York Apple Tours? Most viable political candidates,

Republican and Democrat alike, currently support the death penalty for people. Is it possible that the leaders of the major parties can be persuaded that we need the death penalty for corporations? After all, there are corporations which, on the strength of the evidence, seem more deserving of lethal injection than some residents of Texas's Death Row.

But the death penalty should never be imposed lightly, even for a corporate entity whose life is worth, in Justice Finch's words, "less than that of the humblest citizen." Innocent people's jobs are at stake; creditors may lose significant amounts of money when the corporation is shut down. Also, the act of closing certain kinds of manufacturing plants, mines, and waste depots can generate new risks to the environment and public health, creating the need for years, maybe centuries, of expensive monitoring and maintenance of the abandoned facilities. Nevertheless, some corporations have histories so depraved and products so deadly that the public welfare would seem to demand the death penalty.

In a country ruled by laws, matters of life and death are resolved in the courts, after legislatures have defined the crimes and penalties. What would a trial of a corporation accused of capital crimes be like? Once the trial has started, certain critical questions will inevitably be raised. Did the highest executives of the company know that their products were damaging people and the environment, and did they conceal that knowledge? The tobacco industry—which has given us a preview of such trials—has been shown to have knowingly and covertly laced its cigarettes with dangerous, addictive chemicals. Similarly, in August 2000, the president of Mitsubishi Motors acknowledged that his company had been concealing customer complaints about faulty fuel tanks, clutches, crankshafts, and brakes since 1977. The tobacco industry's actions have resulted in thousands of deaths, but in the case of Mitsubishi there was no evidence that the defects had caused any serious or fatal accidents.[6]

Meanwhile, it was revealed in August and September 2000 that defective Bridgestone-Firestone tires on Ford Explorers *did* cause deaths—knowledge that both Bridgestone and Ford seem to have covered up, or, at best, ignored.[7] In a trial, would this fatal outcome make a difference? Equally important, would prosecutors be allowed to admit evidence concerning the environmental damage caused by the Explorer's heavy gas consumption, excess exhaust emissions, and off-road capabilities? Corporate violence against people and violence against nature often go together.

A question that is bound to be raised at such a trial would be whether there has been a long history of ruthless abuse of corporate power. Some of the largest chemical companies have been devastating environments and killing people in and out of their factories for many decades. Senior executives of the giant German chemical and pharmaceutical combine, I. G. Farben, were convicted at the Nuremberg Trials in 1948 for complicity in such corporate-sanctioned crimes as "plunder and spoliation" and "slavery and mass murder." The sentences meted out for these crimes were short, and two of the top executives who served time in prison later went on to become CEOs of I. G. successor companies.[8] These successor corporations, among the most powerful in the world, are still flourishing and are still capable of causing harm to people and environments. For example, the Bayer Corporation refused to stop selling the antibiotics known as fluoroquinolones for use in chickens and turkeys—a lucrative market—despite the strong protests of the U.S. Centers for Disease Control and the Food and Drug Administration's Center for Veterinary Medicine. The FDA was concerned because of mounting evidence that feeding the fluoroquinolone antibiotic Baytril to farm animals causes widespread bacterial antibiotic resistance, which can spread to humans, and can make related drugs like Cipro useless for treating dangerous infections. Finally, in 2005, after five years of unsuccessful pleas to Bayer for voluntary removal, the Food and Drug administration ordered the veterinary drug off the market, an unprecedented step.[9]

What about crimes so enormous that they cannot be conclusively proven until after civilization and the biosphere have suffered irreparable damage, and perhaps not even then? For example, multinationals, especially some of the interlinked oil, chemical, and automobile corporations, have been fighting the implementation of meaningful efforts to slow global warming. Meanwhile, the ice in Antarctica and the Arctic permafrost is melting, and oceans are on the rise. But the public, by demanding the corporate products that cause global warming, is equally complicit in this "crime," which is perfectly legal in any event. In such instances, trials would not be justified.

I can think of a combination of measures that might, unlike trials, prevent the worst corporate abuse before it happens. (1) Return to the idea of corporate charters that have a fixed time period, say twenty years, after which they require renewal (like radio and television broadcasting licenses), pending a searching review of the corporation's activities. (2) Eliminate, probably by constitutional amendment, the special privileges that corporations have gradually gained in courts during the past 150 years. (3) Close the legal loopholes that enable senior executives to dissociate themselves from the misdeeds of the companies they control. (4) Make it much harder for corporations to evade punishment by jettisoning divisions, changing names, merging with other corporations, or otherwise altering their identities. (5) Renegotiate world trade treaties to restrict the ability of multinationals to use world trade regulations to nullify national environmental and human safety laws. (6) Protect communities by limiting the rights of multinationals to transfer factory operations and large blocks of capital from country to country or state to state without warning and without evaluation of local impact. (7) Change the laws and regulations that allow the largest corporations to avoid paying their fair share of the taxes needed to support the people and environment of the country. (8) Reflect on our own complicity in corporate violence; avoid purchasing products that we don't need and which are socially and environmentally damaging. Utopian suggestions? Not necessarily. Not if we put our minds and hearts to it.[10]

If human rights and a healthy environment are to make it through this twenty-first century, we had better do something soon to reestablish the authority of law over giant corporations. It will be an uphill battle, considering the trouble it took to shut down tiny Apple Tours. But the times are changing; a brighter light is shining on corporate boardrooms and down the corridors of power, even as the unwieldy size, inertia, and Byzantine structure of the multinationals make them vulnerable to unexpected challenges from an angry public and a balky environment. It is too late, of course, for Randolph Walker and the numberless other victims of past corporate crimes. All we can do for them—and for posterity—is promise not to forget.

# FOUR

---

## RELATING TO NATURE IN A HUMAN-DOMINATED WORLD

In relation to the earth, we have been autistic for centuries. Only now have we begun to listen with some attention and with a willingness to respond to the earth's demands that we cease our industrial assault, that we abandon our inner rage against the conditions of our earthly existence, that we renew our human participation in the grand liturgy of the universe.

Thomas Berry, *The Dream of the Earth*[1]

# WILDERNESS AS TEACHER

First fragment. I am in my apartment, a modern, starkly decorated eyrie in a high-rise building. I know it is my apartment in the intuitive but unchallengeable way one knows things in dreams. Actually, I am in a short hallway within the apartment—a parqueted floor and a window at the end of the hall stick in my mind. There is a storm going on, but the other side of the window seems very far away. It grows dark outside; lightning flashes intermittently. Sam, my youngest child, warns me to stay away from the window; he points out that there is a danger of tornadoes. As if to underscore the seriousness of the threat, the lights of the apartment begin to flicker.

Second fragment. My wife, Joan, meets me at the door to the apartment. She looks upset.

"The vine cleaners are here," she whispers. It is clear to me that she wants to send them away—we don't need vine cleaners for our potted plants—but it is too late. Looking past Joan, I see that the vine cleaners, an immigrant couple, are already at work, spraying and polishing the leaves. I can smell the perfumed chemical cleanser they are using. Perhaps, I think to myself, this is a free service provided by the management of the apartment building; but no, it isn't, for now they are done and the man tells me that the vine cleaning costs $8.50. I hand him a $10 bill and get a one and two quarters in change. The couple waits. Do I tip them? A quarter each is hardly enough. Is a dollar too much? I hesitate. Outside, the storm rages. Then the lights flicker again and go out, freezing the tableau.

I opened my eyes with mixed feelings of relief and disorientation. Relief because the vine cleaners had vanished, along with the question of how much to tip them—there is no need to tip people in dreams.

Disorientation because the place I had awakened in was so utterly different from the clinging aura of the dream, and yet this real place was also
for a moment strange and unfamiliar. For I was not in my first-floor bedroom in my house in New Jersey; I was, I gradually remembered,
wrapped in a sleeping bag on a camp bunk in a one-room log cabin on
an island in Lake Temagami, in central Ontario, Canada. The sun was
shining and I could see the sparkle of the water through the birches
outside the window.

It was the end of a five-day family canoeing trip, and we had spent
the last night back at Northwaters, the wilderness canoeing camp whose
gear and canoes we had used. We had started in mid-August, after the
close of the regular camp season, so that our daughter Kate, then a trip
leader at Northwaters, could be our guide. Our son Jon, at that time
twelve years old, had just finished a seven-week session at Langskib, the
brother camp farther up the lake, and he came along too, eager for
more, even after several hundred miles of canoeing that summer. With
Joan, Sam, and me, we made a party of five—Jane, our other daughter,
was busy being a camp counselor for prekindergarten children in New
Jersey, and couldn't be with us.

Kate had just returned from coleading one of Northwaters' wilderness expeditions, an environmental leadership training trip for ten
young men and women down northern Manitoba's Seal River to Hudson Bay. Listening to her stories about forest fires raging along the
banks upriver, polar bears prowling around the campsites farther
down, and the early morning rendezvous in foggy Hudson Bay with
the fishing boat that would take them to Churchill, I had had no
qualms about going out on our thirty-mile lake paddle. But our trip
also had had its moments, including the roaring wind and big waves on
Lake Obobika.

Kate in action was a revelation, whether she was carrying a seventy-
pound canoe with a thirty-pound Duluth pack on her back at a fast clip
along a half-mile portage (she is five feet three inches tall, and slender),
deciding when it was safe for us to be on the water and when it was not,
reassuring nervous parents, baking cakes and bannocks in a reflector
oven by the campfire, showing nine-year-old Sam how to care for himself and be helpful, and subtly guiding Jon's ongoing wilderness training
without undermining his self-confidence. Jon was a revelation, too. A

wonderful canoeist, he carried huge loads on the portages and regaled us with an endless flow of jokes and stories. Suddenly mature and knowledgeable, his seven weeks at Langskib had taught him that to be manly is to be generous, resourceful, strong, fun-loving, and affectionate—traits also shown by his very womanly sister. (At this writing, Jon and Sam, now in their twenties, have followed in Kate's footsteps. They have spent summers as highly skilled canoe trip leaders on some of Canada's wildest rivers and on James Bay.)

During the trip, a curious reversal of roles had taken place. Joan and I were accustomed to receiving good advice from Kate; she had been giving it since she was three years old. But up to now we had always been in charge. In the Canadian woods it was different. Here is an example. It was our last portage; we were in the middle of moving our gear when a large group of campers pushed past and piled up their canoes, partially blocking the end of the narrow trail. I spoke angrily about it to one of the campers, a boy of fifteen or sixteen, who reacted with embarrassment and confusion. Then Kate took me aside.

"What were you trying to accomplish by talking to that boy?" she asked me in a quiet, stern voice. I couldn't reply, partly because I was ashamed of the answer, and partly because it was painful to be reprimanded by my college-aged daughter. A little later, I heard her chatting with the other group's counselor, out of earshot of the campers.

"Because you aren't from this area," Kate said, "you probably don't know that it's a Lake Temagami custom to wait at a portage until the previous group finishes." The counselor expressed surprise and thanked Kate for the information, saying that it wouldn't happen again. And I began to understand a little better that it takes more than canoeing skills to make a good trip leader.

Lying on my bunk on this last morning of our short vacation, I thought of the things we had seen and done on the trip. I thought of the great, freestanding pillar of stone, sacred to the native peoples, that is known as Conjuring Rock. I thought of the blueberries, last of the season, at our Lake Obobika campsite. I thought of the old-growth stand of giant white pines with the small decorated parcels of birch bark and twigs placed carefully against the trunks of the largest trees by unseen hands, probably Ojibway. I thought of the big cat (I was sure it was a lynx; everyone else said bobcat) that swam across the bows of our two canoes, then, gaining the shore, turned calmly to look at us before bounding away. I thought of my fear turning magically into joy as I headed the canoe into Obobika's whitecapped waves.

Only half-awake, I also thought about what David Knudsen, the founder of Langskib and Northwaters, was achieving: the use of wilderness to teach each child how to be fully human, a mensch, a person of honor, integrity, and compassionate judgment. Nor was the teaching only for healthy kids. A year earlier, Kate had led an adult women's group on a highly successful trip. Other counselors guided trips for handicapped children, men's groups, business executives, even military personnel, but the lessons of cooperation, decency, and love and awareness of the environment were always the same.

But what of the vine cleaners? I couldn't get them out of my head. How did they fit in here, on the shores of Lake Temagami? It was my dream; nobody had imposed it on me. What was I telling myself? Was it an anti-city message? I didn't think so; I like cities and know their value. The dream wasn't about cities; what was it about?

The vivid contrast between the life in the dream and the simple cabin where my family was sleeping held at least part of the answer. How can we remain human in a totally contrived world, where our days are cluttered with tasks whose function is to shore up and touch up the illusion we live in? I have seen the brilliant flowers of scarlet runner beans—quintessential vegetable crop of northern cities—lovingly tended by tenement dwellers in Manhattan's Bowery district, but I have also seen employees of a plant service company, on their monthly rounds to skyscraper lobbies and elegant apartments, replacing dying rubber trees in fancy flowerpots with fresh rented plants. My vine cleaners belonged to this latter group, touching up the illusion with their absurd spraying and polishing.

The fact that I made the vine cleaners an immigrant couple was not lost on me, either. When we fill our lives with what the late economist E. F. Schumacher described as bad work, we invite exploitation and abuse of others, especially those who are most vulnerable.[1] The awkwardness about the tip was another way of saying how hard it is to deal with people when we know that the work we have imposed on them is trivial and unnecessary. This applies not only to our employees but also to our friends and to our children. Bad work comes, in part, from being cut off from nature. In wilderness it is easy to see what matters in our lives and what doesn't. But we don't need to be in wilderness all the time to remember what matters; contact with the natural world can be maintained wherever we dwell.

In the dream, a storm was raging outside the apartment; there were tornadoes. The lights were flickering, as if to show me the final dependence and fragility of my constructed environment. One of the consequences of our estrangement from nature is the belief that environmental events need not touch our lives. I did not even have the sense (or strength?) to move away from the window; Sam, only nine, had to urge me to step back. At the end of the dream, the lights in the building went for good, leaving Joan, Sam, the vine cleaners, and me helpless in the dark.

We left Temagami that morning, heading back to suburban New Jersey. The taste of the dream lingered as I dressed and packed, as I said good-bye to the staff, and as I walked to the dock. It lingered as we waited for Jon and his counselor, Dave Torres, to give each other a farewell hug. It lingered as the island slid away behind us. And it lingers still as I live and go about my daily business.

---

## AN OPPOSING VIEW OF NATURE

My old college housemate Claude had dropped by for one of his stimulating visits, an infrequent happening even though his apartment on Manhattan's Upper West Side was only an hour away by car. In college, we had been utter opposites in interests and temperaments—Claude's direction was art criticism, which he celebrated with an élan that gained him a wide circle of friends. Mine was biology, and shyness was my distinguishing social feature. A shared liking for history and the attraction of very different personalities formed a narrow bridge between us, enough to have kept us in occasional contact over the years.

We had been discussing European politics and the question of decentralism over a light lunch on the patio, and then, in one of those curious shifts that conversations take, we had abruptly changed the subject. He was holding forth in his usual compelling fashion.

"The trouble with nature writing," Claude said, "is that it's so artistically accepting, so uncritical. Lopez, Abbey, and their ilk—why civilization seems to have passed them by without leaving a mark. Culture implies the ability to discriminate, to judge—not sometimes, all of the time. Why should nature be exempt from criticism?"

He readjusted his long, elegantly trousered legs carefully on the lounge chair, glared briefly across the patio at a song sparrow sitting on a dead branch of my Cortland apple tree, and continued. "I fail to comprehend why you nature-lovers don't hold nature to the same kind of exacting standard that a cultured person applies to a bottle of Margaux or a performance of *Rosenkavalier*. Undiscriminating adulation is barbaric."

Just then a starling arrived at the bird feeder, displacing a blue jay. "Do you see that?" cried Claude. "That is exactly what I mean."

"Too aggressive?"

"No, no, that's not it at all. Look at the preposterous thing. Surely you can see what I'm referring to?"

"All I see is a starling. It seems healthy."

"It seems healthy!" Claude stared at me in disbelief. "You're not understanding me. Try to look at it as if it were a painting, a painting you did in beginner's art class. I am your instructor, and we are examining it together. Now what's wrong with it?"

"I don't see anything wrong with it. It looks just like a starling."

Claude shook his head, closed his eyes for a moment, then cut himself a small wedge of brie from the round that was on the little table at his elbow.

"I apologize," he said after swallowing the morsel of cheese, "I was asking too much. Let's start with the head. What do you notice about the head?"

"It's a different starling," I observed, "the first one flew away."

"Never mind about the first one! I don't give a damn if it's the first one, the second one, or the fiftieth. They're all the same. Don't you notice that its head is all wrong? It's not the right size for the rest of its body. If Michelangelo's *David* had been proportioned like that, they would have cut it up for souvenir paperweights years ago. And that garish yellow beak—badly colored and crudely shaped. No decent design studio would tolerate that sort of work."

I saw what he was getting at, and nodded. Encouraged, he went on. "About the absurd shape of the bird, there is no need to comment. But I must say something about the feathers. Those speckles. That cheap, iridescent shine—like a $79 suit in a sleazy men's discount store. Aesthetically, the bird doesn't work at all."

"They're very successful. There are millions of them in North America."

"Common," sniffed Claude. "Their numbers don't impress me."

I tried to be conciliatory. "You have to remember that they're an exotic species. They don't belong here. Somebody brought a few over from England in the late nineteenth century and let them loose in Central Park in New York because they were mentioned by Shakespeare."

"That's quite irrelevant. True aesthetic sensibility is universal. It doesn't recognize national boundaries. I'm sure they are just as hideous in England."

"I don't think that Shakespeare ever said anything terribly negative about how starlings looked."

"Shakespeare had more important things on his mind: ungrateful daughters, jealous husbands, tyrannical usurpers, malicious servants. If he had had time, I have no doubt that he would have been severely critical of starlings. 'That ill-visag'd fowl.' Perhaps he might have worded it slightly differently."

I tried to come up with some ersatz Shakespeare to make the case for starlings—even though I don't much like them myself—but I couldn't compete with Claude. Still trying, I went inside to the kitchen to fetch glasses of iced tea and brought them back out to the patio. For a while we spoke about our mutual friends from college, what they were doing, how much they had aged. The song sparrow had been trilling loudly during this part of our conversation; Claude broke off his remarks and made a face.

"What a racket. That brings us right back to what I was saying earlier. If that were an instrumental piece, the critics would unanimously call attention to its total failure to explore musical space, its slavish commitment to melody, its inability to expand into another temporality. But because it's a bird, a privileged part of nature, it can commit a cultural felony and get off scot-free. And look at the grubby brute. What has it been doing to its chest? Has it been carrying a leaky fountain pen?"

"Every song sparrow is like that," I protested. "The dark spot on the breast is characteristic, and . . ."

"You don't seem to understand. Just because they *all* have it is not an excuse. But let's get off the subject of birds. I can see that you're looking at them through rose-tinted binoculars." He paused briefly to let me appreciate the joke. "I'm trying to make a general point. We have re-made the world from top to bottom. Our canals cut from one ocean to the next. If there is something we want inside a mountain, we remove the mountain. If we need an offshore island where there isn't one, we make an offshore island. If a forest has been used up, we plant another in its place."

"Actually . . ." I interposed, but Claude held up his hand.

"Let me finish. The world is now largely a product of our own making. And like any artist or designer, we take the credit or blame for what we produce. If its form and texture are attractive and pleasing, we are deservedly praised; if it is ugly and tasteless, you can be sure that we will be harshly criticized. Why shouldn't nature be held accountable as well?"

"Actually, those planted forests don't work very well. I mean often they don't even grow, like the attempts to replant the cut-over areas of the Cascades in Oregon. In some places they tried and failed three times. And when the planted forests do grow—like the slash pine forests in Florida and Georgia, or the eucalyptus forests in China or Israel—they're very monotonous and sterile."

"Then don't plant them. Didn't you once tell me that the English moorlands used to be forested? Think how many poets and novelists have written about their wild beauty. Obviously they're much better off the way the early Britons left them: without trees. They had excellent taste, the Celts or the Saxons, or whatever they were."

"Are you suggesting that the deforestation of the moorlands was a planned and deliberate creative act, like building Stonehenge?"

"Why not? We don't have any way of knowing what those folks thought. They may have been very artistically perceptive. If the forest was offensive, why shouldn't they have removed it for that reason and replaced it with something nicer?" Claude broke off his remarks, got up, stretched, and walked over to the patch of lawn that borders my vegetable garden. "Your little bit of moorland," he said, smiling.

I wasn't going to be baited. "Claude, you seem to believe that nature is just another collection of objects to be judged, like pictures in an exhibition. Do you really think that our aesthetic sense exists completely apart from nature? I've always understood that our idea of the beautiful is derived from natural forms in the first place. The Japanese provide examples. Their sand gardens, raked into patterns like ripples on a lake or wavy lines of surf on a beach. Their bonsai, imitating ancient, wind-pruned trees on a rugged mountainside . . ."

"Their Noh performances. Their tea ceremonies. Oh, come now. You're confusing nature nostalgia with aesthetics. I tell you humanity has expanded beyond nature. The most highly developed appreciation of the beautiful in art that I have ever encountered belongs to an urban friend of mine who once told me that he never left Manhattan until he was eighteen, when he went on a day hike with some college buddies in Harriman Park. He came home an hour and a half early because he couldn't stand it. But if I wanted a sensitive evaluation of a Japanese landscape painting I was thinking of buying, his opinion is the only one I would trust without question. Do you think he acquired this superb ability by looking at those nasty trees along the sidewalk on East 64th Street? Or maybe you think there was an influential potted rubber plant in his parents' apartment when he was a child?"

"Claude, haven't you ever enjoyed nature, been in a natural environment that you could savor, that made you feel good?" I asked. "I mean, made you feel good in a way quite different than you would if you took a trip to the Louvre?"

"Certainly I have," Claude answered. "I went to the Grand Canyon last year. It was quite nice. And some parts of it were really well composed, with a play of light, color, and line that was nearly the equal of what you might find in an early Daumier. Of course, Daumier didn't bother much with landscapes.

"Mind you," he continued, "the Grand Canyon is not beyond criticism. Its scale is too big to allow it to make any kind of modulated or nuclear artistic statement. That's a problem, but if you are willing to restrict your view while looking down from the rim, not try to take it all in like the silly tourists swiveling their heads from side to side, then it's possible to have a pleasant experience."

"Was that the only place you went on your trip?"

"No. I traveled about other parts of Arizona—my friends insisted on it. I saw the giant meteor crater and then made my way south."

"What did you think of the meteor crater?"

"Well, you know it wasn't nearly as attractive as that open-pit copper mine near Bisbee. They're both craters, I grant you, big round holes in the ground, but I found the lines of the Bisbee mine, with those striking terraces around the sides, much more satisfying. That's a case where nature might have done better by imitating art."

I was struck by the thought that a copper mine, which some view as an environmental blight, was for Claude a thing of beauty. Clearly, natural context was of no importance to him. How could I explain this? I decided not to try.

For the last few minutes, Claude had been mopping his face with a paper napkin. "I think we ought to go inside," he said; "it will be cooler there."

"Only a little cooler. The electric company has imposed a brownout. Too many people using electricity during this Labor Day heat wave. I'm afraid to turn on my air conditioner—it might be damaged by the low voltage."

"I don't see why they don't build more power stations if they can't supply enough power," Claude responded irritably. "It's a perfectly simple engineering problem. Well, I think I'll drive back to the city. The air conditioner in my car is working, at any rate. Many thanks for your hospitality; I enjoy our visits."

"I'm sorry about the air conditioner, but nature does impose some limits," I said. "We can't just change the whole climate of the earth if we're too hot."

Claude laughed and patted me on the shoulder. "There you go about nature again. You really are an incurable romantic. We made the world hot with global warming, as you yourself pointed out to me the last time we got together. Now we'll just have to cool it off again. We'll find a way. Bye."

There was no chance for me to respond. Claude was already striding toward his car, which was parked at the far end of the block. His earphones were in place, and the iPod fastened to his belt was switched on. I turned away. Then, as I walked back to the house musing about our conversation, I suddenly realized just how Shakespeare would have phrased a complimentary remark about starlings, if he had had the time. I would tell it to Claude during his next visit.

## DEATH OF A PLASTIC PALM

All over the United States, phone companies have planted artificial trees. An old-growth, imitation pine went up a few years ago in the community of Franklin Lakes, New Jersey, where it towers one hundred feet over barns and fields. Holes in its synthetic brown bark admit bundles of black cables that travel up through the hollow metal trunk to the wireless digital phone antennas hidden among the plastic pine branches high above. In the desert near Phoenix, Arizona, where pines are scarce, another phone company conceals its antennas inside a giant, imitation saguaro cactus. Evidently some large corporations are finally beginning to realize that the destruction of nature by their business activities is not acceptable. To improve public relations, they are now providing facsimiles of the nature they eliminate. This solves many annoying problems, but, alas, other difficulties have arisen. Copying nature is not as easy as it looks—it's not like using a photocopier. If the phone companies don't know this, somebody ought to tell them.

I first became aware of the downside of artificial nature when I was living in Middlesex, New Jersey. To get to work at the university I had to drive through South Bound Brook, a down-at-the-heels little town which was then nestled cozily between American Cyanamid's huge organic chemical factory on one side and GAF and Union Carbide— also cooking up organics—on the other. True, the Raritan River runs alongside of South Bound Brook, and George Washington, an experienced surveyor who had seen many rivers, once described the Raritan as the most beautiful of all. But the river's magic seemed to work feebly in South Bound Brook; my memories are only of floods, brown mud, and the occasional corpse of carp or sucker floating belly up past the dirty red brick of the aged factory walls.

Then one day as I made my daily commute, I noticed an unexpected splash of color in the town. A popular bar along the road I traveled had erected a very large plastic palm tree on the sidewalk next to
the entrance. Its ornate fronds swayed gently in the torpid breeze,
beckoning to the factory workers passing by. If the green of the leaves
was a shade more vivid than one generally finds in nature, and if the
bark was a trifle more orange, nevertheless the palm was clearly making an important statement. From then on, every morning on the way
to work and every evening coming home I looked at that palm, with
feelings one could liken to those of John Bartram and his artistic son
Billy as they gazed upon their great discovery, the elegant, flowering
*Franklinia*, growing alongside the Altamaha River in southern Georgia
back in 1765. The plastic palm became a fixture for me, a necessary
part of the landscape. It replaced the shaggy-barked, curving river
birch or massive, buttressed sycamore or pin oak that I might have seen
in South Bound Brook had I made my trips in the time of the Bartrams
or George Washington.

Months passed, perhaps a year and a half, before I noticed one day
that something was wrong with the plastic palm. The leaves were yellowing. The change was slight, observable only at the tips of the fronds, but to
one who is keenly aware of nature, there could be no doubt. Although
I am not an expert on plastic, I do know something about real plants.
When the leaves of an evergreen tree like a pine or palm start to lose
their chlorophyll it is a sure sign of illness—ask any plant pathologist.
During the next few weeks I cheered myself with the hope that the palm
was only suffering from a transient nutrient imbalance. Maybe it will
recover, I thought. Give it some time.

But its condition only grew worse. The yellowing spread, covering all
the leaves; soon the friendly, garish green was just a memory. Then the
broad leaf blades began to disintegrate and fall away, each leaving
behind its central, supporting rachis. By autumn, not a single frond was
left intact. The once-leafy crown of the palm was now a forlorn collection of bare plastic (or aluminum?) wires drooping out of the top of the
trunk. It reminded me of a coconut palm that used to grow on the
beach in Tortuguero, Costa Rica, in front of our research station. On
an August morning in the mid-1960s, when I was a graduate student,

a group of three or four CIA-supported anti-Castro freedom fighters, who used our little airstrip on the beach to fly in toilet paper and other supplies for their camp upriver, shot away most of the leaves of the coconut with machine guns. We never knew why they did it—perhaps the tree had offended them in some way—but whatever their motives, when they were finished it looked a lot like the plastic palm in South Bound Brook during its last few months.

I could no longer bear to see the afflicted palm and averted my head when driving past, traffic permitting. At least a year went by until the saloon owner had the remains removed. Just before it went, I noticed out of the corner of my eye that the bark was splitting and peeling off in tattered strips.

Eventually I forgot about the plastic palm, but was reminded of it one day when I saw another one in front of a store on Route 36, on the way to the ocean beach at Sandy Hook. I make that trip often, taking students on field trips, and each time I looked carefully at Palm II, as I called it. Sure enough, after a period of months, I noticed the now familiar signs of yellowing at the tips of the fronds. It progressed slowly, but I knew what the outcome would be. Palm II was dying. This time, instead of grief I felt a thrill of excitement: the death of the South Bound Brook palm had not been a special case. I was the first scientist to discover a disease of imitation plants.

What is causing this apparent epidemic among plastic palms? Not knowing where else to turn, I consulted the fountainhead of information about tree diseases, *Pirone's Tree Maintenance*. Going directly to the section on palms, I soon came across *Pythium* root rot, which "is accompanied by yellowing and wilting of the leaves one after another, until the bud falls from the top of the plant." Admittedly, there is some similarity here, but I don't really think that the plastic palm wilt (this is what I have named my new disease) is fungal, like *Pythium*. Air pollution, which has been known to affect and kill plants for 150 years, is a more plausible culprit. Sulfur dioxide, fluorides, hydrocarbons, ozone, peroxyacetyl nitrates, nitrogen oxides, ammonia, and assorted acids are all toxic to plants, and all are common enough in urban and even in some rural air. A glance at Richard Harris's important book, *Arboriculture*, confirmed my suspicions: "Leaves are the plant parts most likely to show symptoms of air pollution injury," says Harris, and I have no reason to doubt his word. Perhaps the plastic palm wilt will turn out to be caused by air pollution.[1]

The implications of my discovery are disturbing. After all, why take such pains to imitate nature if we are going to have this kind of trouble with its replacements? What went wrong, I wondered? Surely the original idea of reproducing only the brightest and most beautiful inventions of nature—green leaves, waving branches, stately tree trunks—was a sound one. There must be a catch, I realized. Could it be that nature packages its inventions? Do the parts belong to an elaborate, interlinked system that can't be completely disassembled? Of course, that's it. Pull out the green leaves and waving branches for copying, and like it or not along come some other pieces of nature: rot, decay, even a pretty good facsimile of mortality. Wow! I thought, is the phone company aware of this? If there can be a plastic palm wilt, why not a plastic pine blight or a steel cactus canker?

Then an even more worrisome idea occurred to me. When we copy nature, not only do we get things we don't want, but some critical parts are always missing, parts that we can't copy, the best parts of all. These are the signature processes of nature: self-repair, regeneration, and rebirth. Nature holds on to them tightly; they are not just copyrighted, they are copyproof. Decay and death are tolerable, even beautiful, in nature because they are food for regeneration and rebirth. In imitation nature, decay and death lead nowhere—uncoupled from the grand cycle, they are shabby, pointless, and unredeemably grim.

It almost seems as if nature does not like synthetic imitations. Imitation may be the sincerest form of flattery, but nature is neither flattered nor amused. Her anger is a serious matter. That's why I am concerned about the one-hundred-foot plastic and metal pine in Franklin Lakes. The machine-gunned coconut in Tortuguero eventually put out new leaves, but what will happen when the imitation pine loses its branches in a high wind or ice storm, as it inevitably must? Will new branches sprout from the trunk; will the broken giant be ringed by imitation pine saplings reaching upward in the sudden sunlight to replace it? No, the phone company will have to pay to have their pine fixed. And pay and pay and pay and pay forever for what once was free.[2]

## SCIENTIFIC DISCOVERIES AND NATURE'S MYSTERIES

One of the joys of science is its steady production of wonderful discoveries. At irregular but not intolerable intervals, it throws out to the waiting world findings that add spice to a thousand morning newspapers and enliven the conversations of countless barbers and taxi drivers. Some branches of science are especially fruitful in their production of wonders: physics doles out an inexhaustible supply of elemental particles with clever, memorable names like *quark*. Medicine provides powerful antibiotics freshly isolated from weird sources such as exotic muds and frog skin, arraying them against an equally fresh crop of antibiotic-resistant microbes. Geology reveals previously unsuspected earthquake faults under major cities and dazzles us with hidden, underwater chains of volcanoes in secret, violent eruption. Even paleontology can be counted on for the latest, well-documented theory of how birds came to fly or, as I have already discussed, why the dinosaurs became extinct. And then, every ten years or so, like the return of a speeded-up Halley's Comet, animal behavior brightens the scientific horizon with a new and incredible explanation of how animals find their way home.

Take, for example, the discovery by Peter Berthold and three colleagues that bird migration has a rapidly evolving genetic basis.[1] Evidently, the inherited tendency of a population to choose a particular migratory direction can change in just a few years. Berthold, working in Radolfzell in southwest Germany in the 1990s, made the most of an earlier observation: since the 1950s, increasing numbers of blackcaps, Old World warblers, have been spending the winter in Britain, traveling from their breeding grounds in Germany and Austria during the fall migration. Traditional wintering grounds are in the western Mediterranean countries far to the south; most blackcaps still go there in the fall. But the blackcaps wintering in Britain have done well; global warming (or chance) has ameliorated the winter climate, and the British are dedicated bird feeders, so there was no shortage of food.

What Berthold did was to capture forty male and female blackcaps in England, transport them to Germany, breed them, and raise their young. Another group of nestling blackcaps from nests around Radolfzell, presumed to be mostly or entirely from the Mediterranean wintering population, served as a control. In the fall, when the young birds of both groups were fully fledged, they began to show migratory restlessness and to point in the direction they wanted to fly, a phenomenon noted many years earlier by the ornithologist Gustav Kramer. Berthold put them separately in special, outdoor, circular cages that recorded which way the birds were facing. The British-bred birds pointed slightly north of west, toward Britain; the Radolfzell controls faced southwest, toward Spain.

There are a few flaws in the experiment, as there are in all experiments: for example, it would have been better to supplement the Radolfzell controls with a group of captive-bred nestlings whose parents were known to winter in Spain. But no need to quibble. This experiment reinforces the older demonstration that the direction birds face when they want to migrate, probably the same as the actual departure heading, is inherited. More exciting, Berthold has shown that natural selection can quickly modify the inheritance of departure heading in migratory birds. In less than forty years, the proportion of German-born blackcaps that overwinter in Britain went from zero to around 10 percent.

Did Berthold's experiment finally answer the ancient question of how animals find their way home during long journeys? Hardly. Getting the starting direction right does not necessarily guide a bird through a long and complex migration. Like the pot of gold at the end of the rainbow, the definitive and final answers to the mystery of animal navigation are always over the top of the next hill. Looking back at a succession of great discoveries about animal navigation during the past half-century, I see a clear pattern, a regular sequence of events. It goes like this: discovery, scientific enthusiasm, public awareness and excitement, supportive findings, complications, disillusionment, new discoveries. Notice that I use the word *complications*, not *disproof*. Major discoveries concerning animal navigation (and other natural phenomena) rarely get disproved, just complicated.

I should describe what I mean by navigation. This is the ability, shared by many species of birds, sea turtles, certain butterflies, various fish, and a host of other species, possibly including dogs, cats, and people, to find a particular place that is far away, while traveling over unfamiliar territory. True navigation is more than orientation to known and fixed landmarks. To a green turtle swimming in the South Atlantic, out of sight of land and far above the ocean floor, heading for Ascension Island, a five-mile speck of land a thousand miles ahead, there are no fixed landmarks—at least, not ones that are evident to us.

Nor does a mere "compass sense" suffice for navigation. A compass sense, being able to tell north from south from east from west, is probably shared by most species of animals, including invertebrates. It is a necessary part of navigation systems, but usually it can't do the job by itself. You can use a compass to find home only if you know what direction your home is compared with your present location, which requires the equivalent of a map and knowledge of where you are on that map, the "X—you are here" of hiking trail and shopping mall signs. I can describe another experiment that should make this clear.

More than a decade before Berthold did his work, a Dutch scientist, A. C. Perdeck, performed an even more spectacular research feat, with remarkable results.[2] Each year in the fall, starlings that breed in the spring and summer in the eastern Baltic countries fly southwest in large flocks, traveling across the Netherlands on their way to their winter territory in southern England and Ireland and northern France. They fly in mixed flocks, with the newly fledged, inexperienced migrants accompanying adult birds who have made the trip before. Perdeck captured eleven thousand of these migrating starlings near the Hague and carried them to Switzerland, far to the south of their familiar migratory route. There he banded them and released them, but he took care to release them in separate flocks composed of only adult or only immature birds.

Recoveries of banded birds revealed a striking difference: the immature, first-time migrants simply continued on the typical southwestern fall compass heading (as if they were still in the Netherlands) and ended up in southern France and Spain where they spent the winter, hundreds of miles south of their traditional winter habitat. But the adults flew in an unusual northwestern direction over unfamiliar territory, landing in their regular wintering grounds in northern France and, in a few cases, southern England. The adults showed true navigational ability; the juveniles merely had one component of that ability, compass sense.

The incredible conclusion to this experiment was that the following spring some of the younger but now mature birds returned to their place of birth in the eastern Baltic countries to nest. And the next fall, some of these birds migrated again—not to England, Ireland, and northern France with all the other starlings—but to southern France and Spain, where they mistakenly had spent their first winter.

If I have explained it right and if you have followed everything so far, you may have noticed that the results of the Perdeck experiment complicate the interpretation of the results of Berthold's later study. Perdeck showed that although an innate compass heading (including, by implication, the initial compass heading of migratory restlessness) is a factor in bird navigation, it is not really necessary; the adult starlings in his study abandoned it altogether when they were displaced off course to Switzerland, an unfamiliar location. The inborn compass heading was followed only if the adult birds *already knew* they were flying in the right direction.

The history of research in animal navigation has many other incredible discoveries and theories, always followed (or preceded) by equally incredible complications. I will quickly describe some more examples.

Birds and honeybees have an accurate sun compass based on visual detection of the position of the sun, plus an accurate biological clock to correct for the change in the sun's position with time. But the sun does not have to be visible for the compass to work; any patch of blue sky has a light polarization pattern that gives the sun's location, as Karl von Frisch showed for honeybees. Bees, and probably birds, can see this polarization pattern in the sky; people can too, if they work at it. (The human ability to see polarized light was first described by the Austrian mineralogist Wilhelm Karl von Haidinger, in the 1840s. The so-called Haidinger's brush, a faint yellow bow-tie shape, can be seen by gazing at the white polarized light from the screen of any laptop monitor; orientation of the bow-tie indicates the plane of the polarization.) Yet if the polarization pattern is obscured by total overcast, bees and birds *still* have an accurate compass.

Some species of birds can navigate at night. E. G. Franz Sauer, one of my teachers, first put them in a planetarium and showed that by rotating the night sky he could make caged warblers change their heading in a predictable way. S. T. Emlen extended this work to identify the important features of the night sky for indigo buntings; the birds paid particular attention to the area around Polaris (the North Star) and the

Milky Way. Yet nocturnally migrating birds can hold a true heading while flying in thick clouds on an overcast night when no stars are visible, as radar studies have shown.

Birds were never thought to have much of a sense of smell, but Floriano Papi and his colleagues demonstrated that homing pigeons can use their sense of smell to find their way to the home loft when it is a few miles away.[3] He came up with a simple but brilliant theory to explain how this olfactory guidance system could work, even if the wind is blowing from the wrong direction. Soon, seabirds like petrels and albatrosses were shown to locate squid, fish, and krill in the ocean by their odors.[4] Could familiar smells be the magic migratory cue? Yet pigeons can find the home loft when it is hundreds of miles away, out of range of familiar smells. Moreover, Papi's experiments, which worked well near Florence, Italy, did not work in Ithaca, New York. While disillusionment with the olfactory guidance hypothesis was growing, other researchers were concentrating on a different sensory cue that can provide orientation information, the earth's magnetic fields.

The earth's magnetic field can be detected by compass needles, snails, honeybees, sea turtles, birds, and many other animals. Charles Walcott, James Gould, and J. L. Kirschvink found crystals of magnetite, magnetic iron oxide, in the heads of homing pigeons, indicating a possible mechanism for the birds' detection of magnetic fields. Soon, a spate of papers claiming to show magnetic field navigation in birds, sea turtles, and even lobsters appeared in the scientific journals.[5] Yet it is extremely difficult to get bicoordinate map information (latitude and longitude) from magnetic fields (although compass direction is easy); and there is still no good explanation of how an animal's sensory and nervous system could accurately measure the horizontal and vertical components of magnetic fields, and then turn the incredibly complex data set into a map. By 2006, the tables were turning again: Anna Gagliardo of the University of Pisa, Italy, tested magnetic sensing and olfaction side by side in homing pigeons and found that inexperienced birds deprived of their magnetic detection ability were able to return fifty kilometers to their home loft, while birds deprived of their sense of smell did not return.[6] No doubt the debate between the olfactory and magnetic schools will continue.

My favorite explanation of animal navigation is inertial guidance, the navigational system of rockets, which received much attention from behavioral scientists around the time of the Apollo moon voyages. Inertial guidance (or "dead reckoning") requires both a map and compass sense; an animal would have to register, remember, and analyze the duration, speed, and direction of every phase of its travels, comparing its position against its internal "map." This was first suggested in a few paragraphs of a letter to *Nature* dated April 3, 1873. The author, Charles Darwin, wrote,

> We must bear in mind that neither a compass, nor the north star, nor any such sign, suffices to guide a man to a particular spot through an intricate country, or through hummocky ice, when many deviations from a straight course are inevitable, unless the deviations are allowed for, or a sort of "dead reckoning" is kept. All men are able to do this to a greater or less degree, and the natives of Siberia apparently to a wonderful extent, though probably in an unconscious manner. . . . Whether animals may not possess the faculty of keeping a dead reckoning of their course in a much more perfect degree than can man . . . I will not attempt to discuss, as I have not sufficient data.[7]

The great naturalist Henry Walter Bates and a companion had their lives saved by a ten-year-old Indian boy who was traveling with them, when they wandered away from the Amazon River, where they had moored their canoe, and became lost in an unfamiliar part of the rain forest. The boy, who had been busy playing with a bow and arrow during their increasingly frantic wanderings, guided them directly back to their canoe after Bates finally thought to ask him for help. "I believe he had noted the course we had taken almost unconsciously," wrote Bates after the adventure.[8] Yet it is hard to imagine such a system of inertial guidance, if that's what it is, working for an arctic tern during its yearly round-trip migration of twenty-five thousand miles, one-tenth the distance to the moon—or, for that matter, for bar-tailed godwits, migratory wading birds, which have been known to fly eleven thousand kilometers without stopping (six days, if the winds are favorable) during their annual migration between Alaska and Australia or New Zealand.[9]

Finally, there were the studies of the Canadian fisheries scientist, C. Groot, who was trying to understand how juvenile sockeye salmon find their way through a complex maze of interconnected lakes to reach the Pacific Ocean. Having shown that they were perfectly capable of orienting themselves using celestial cues, including polarized light, under clear or partly cloudy skies, Groot demonstrated that they did not need these cues to orientate properly under completely cloudy skies or in the laboratory. Perhaps, he thought, inertial guidance was involved—maybe the salmon smolts had memorized all the twists and turns and changes in speed on the way to the enclosed experimental tank. But when he transported them under deep general anesthesia, they oriented as well as they did back in their lakes under a clear sky. In the end, he described their orientation system as "type-X orientation," and left it at that.[10]

And we are left, after a heady succession of discoveries, theories, and complications, with the gleaming, undented core of a great mystery. Is it, as some have suggested, that orientation mechanisms already described, working in tandem, often redundantly, in various combinations, can explain all feats of animal navigation? I doubt it. Are there fundamental phenomena yet to be discovered? Perhaps. Only one thing is sure: the methods of reductionist science skillfully applied over many years have not solved the problem of how animals find their way home. Is it permissible to wonder whether the reductionist approach *can* answer all of the important questions life poses? At any rate, I look forward to years of discovery, disillusionment, and more discoveries in this exhilarating, ever-daunting quest for a deep understanding of one of the most puzzling mysteries of nature.

## I REINVENT AGRICULTURE

After coping during the workday with understanding the mysteries of animal navigation or trying to figure out the best ways to protect an endangered species, I like to vary the intellectual routine by doing something direct and "real" when I get home. Some build cabinets, some sing in choral groups, others join first aid squads. I take care of our garden. Gardening is an especially good activity for eggheads because it combines physical labor with the opportunity to be creative.

I used to think that gardening was a pretty straightforward, even simple, way to pass the time. But my experience a few summers ago changed my mind.

I like to eat what I grow, so I have little patience with flowers and other ornamentals. I tolerate the ones that were planted around the house when we first arrived: yew, hemlock, cherry laurel, forsythia, flame azalea—the usual things provided by botanically challenged, cost-conscious builders. If these woody perennials get sick I do what I can for them, given my lack of knowledge and low level of interest. Sometimes I fertilize, sometimes I continue to refrain from fertilizing, depending on my intuition in the particular case. The same goes for watering. For example, our flame azalea declined over the course of several years to one tiny branch with eleven leaves on it. I didn't know whether I was keeping it alive against all odds, or preventing it from making the full recovery that it would be capable of if cared for by somebody else. I view my role in the treatment of our ornamental plantings like that of the remote HMO clerk who makes the critical decisions in a complex medical case; any unfortunate consequences of my ignorance are not my concern.

Fruits and vegetables are another matter. I am not much more knowledgeable about them than about azaleas, but I try a lot harder. Because this is a story about vegetables, I will skip the fruits, except to note that we pick our own sour cherries, figs, pears, plums, grapes, apricots, and apples—in sharply decreasing order of abundance (squirrels get nearly all of my apples before they ripen). Enjoyable as it is, however, to eat my breakfast cereal with figs from the garden (in New Jersey!) in September and October, it is the vegetables that provide the deepest satisfaction. Not that I like collards and potatoes more than fresh figs and iced sour cherry soup—I don't—but it is in the vegetable garden that I have been able to make my contribution to world agriculture.

Perhaps I should rephrase that. I haven't discovered anything that millions of gardeners and farmers didn't already know. It is not a contribution in that sense. But I discovered it for myself—I didn't read about it in a book or hear about it from another gardener. What I discovered is the secret of feral vegetables. (Some people call them "volunteers." You will soon see why this is much too benign a label.)

Feral, in the way that I am using the word, means having escaped from domestication and become wild. One hears about packs of feral dogs, ownerless, wary of people, living off the land in suburban and rural woods, and occasionally attacking children. Or solitary feral cats indulging their natural love of independence and wreaking havoc among local rodent and bird populations. My first feral vegetable was a patch of coriander, also called cilantro. Originally, I had planted coriander seed purchased from a seed house in Oregon. It grew fairly well, and we ate it in guacamole and curries. But we didn't eat all of it; a few plants flowered and produced seeds that fell on the ground and lay hidden in the soil. The next year, I planted more coriander in a different part of the garden. Several weeks before it came up, I noticed scores of little corianders shooting up near what had been last year's patch and had now become the space between two rows of potatoes. The feral coriander was much more vigorous and luxuriant than the planted stuff: so abundant that it supplied all our requirements for the herb, with enough left over to flower and go to seed again. This happened years ago, and there has been no need to plant coriander again. Although I am never sure exactly where it is going to appear, there is always plenty of it somewhere in the garden. For a deeply lazy person like me, this is an ideal arrangement.

Soon I discovered other annual vegetables that were happy to escape domestication and lead a feral existence. Oriental red mustard greens were the first to follow suit; the feral plants, like the coriander, did much better than the ones I had planted. This, as I found out from repeated experience, is a general rule: things that plant themselves sprout earlier, grow faster, and have less trouble with pests than their planted and managed domestic counterparts. All that is required from me is the ability to recognize the feral vegetables at an early stage and avoid weeding them out, unless they are in the way of something I want more. Gradually, in my mind, a revolutionary new system of agriculture was taking shape; when perfected, all I would have to do is sit on the patio and let the feral vegetables raise themselves.

After the mustard came cherry tomatoes. If I planted their parents in a row next to my snow pea trellis, the following year little tomato plants would appear under the snow peas and be big enough to tie to the trellis in late June, after the snow peas were finished. Then beans joined the list of feral vegetables, plus an occasional potato, and even a pair of hefty collards, which didn't come from seed but from a couple of plants that had been overlooked when the row was cleared the previous fall. The garden started to look terribly disorganized, with miscellaneous clumps of vegetables coming up in unpredictable places; but the feral vegetables were feeding us, and that's what counted. Of course it was too good to last.

Early one summer, a new feral vegetable appeared in the space between two rows of tomatoes. It looked like a little squash plant, which puzzled me because I hadn't had squashes in that spot last year. Then I remembered that we had put down compost in the fall, and it might have included seeds from a supermarket Japanese winter squash of the buttercup (Kabocha) type, our favorite. "Good," I thought, and left it alone. The squash grew rapidly, and I trained the vines out toward the edge of the lawn so that they wouldn't bother the tomatoes. Sure enough, when the first squashes appeared, they had the typical buttercup turban shape; eventually we had three or four tasty meals from them.

Meanwhile, other feral squashes were growing from the same clump. The leaves were different, much larger and silvery green in color; there were sharp prickles on the leaves and stems. They were rowdy vegetables: in no time at all they had leaped over the Rose, Aunt Tettrazine's, and Tommy Toe tomato plants and were heading south toward the Kennebec potatoes. "Good. More squash," I said to myself. I gingerly picked off a few of the gigantic leaves that were shading the tomatoes and left, with my wife Joan, on a two-week trip to Glacier Park, Hell's Canyon, and Spokane.

When we got back in mid-August, the tomatoes and potatoes were gone. At least I couldn't see them under a billowing sea of squash leaves. To get into the garden I had to step on a meshwork of squash stems, which didn't bother them because each had the tensile strength of one of the cables holding up the George Washington Bridge, and about the same thickness. Then I saw it—a monstrous squash. It was long and yellow-green in color, with dark green stripes, a bit like a zucchini with elephantiasis. Looking around, I noticed others resting heavily on the potato hills or peering out from the mixed vegetation where I assumed the tomatoes still were.

What variety of squash was it, I wondered? Protected by cotton gloves, I cut a length of stem containing both leaves and a clublike fruit, threw it in the back of the station wagon, and drove to Oved Shifriss's house. I have mentioned Professor Shifriss before: at that time he was long-retired from a distinguished career as a plant breeder. As I noted, his development of the Burpee Big Boy tomato in 1949 helped to launch America's passion for home vegetable gardening. More important for me, he knew everything there was to know about squashes. He was resting when I arrived and seemed taken aback when I dumped my squash on the top of one of his filing cabinets.

"Have you ever seen anything like this?" I asked.

"No," he replied in a tone that told me he would be happy if he never saw it again.

"What is it?"

"All I can say is that it's in the genus *Cucurbita*, and probably has genes from two very different species."

"Can I eat it?"

"I wouldn't," said Oved, politely indicating that I should take my squash with me when I left.

I figured out what had happened. The Japanese squash whose seeds we had put in the compost must have been a hybrid. Most of the off-spring of hybrid plants do not have the appearance and favorable characteristics of their parents. The reason for this has to do with chromosomes. Generally, chromosomes come in pairs (or multiples of pairs), and in a long-established variety, each chromosome in a pair is genetically much like the other. When two varieties are crossed to form a hybrid, the offspring receives one chromosome from each of the many pairs belonging to each parent. Therefore the hybrid predictably gets half its chromosomes from one parent and half from the other; and all the first-generation plants from that cross are similar to each other, although not to their parents. But when two hybrids are crossed—resulting in the second-generation seeds from my hybrid squash—a vast number of combinations of chromosomes are possible. Some second-generation hybrids, like the buttercup squashes we ate, are similar to their hybrid parents; most are not.

This means that if you want to grow the same hybrid vegetable again next season, you shouldn't save seed from your crop. Hybrid vegetables are favored by the big, modern seed companies, because you have to buy new hybrid seeds every year. Agribusiness loves dependent gardeners and farmers; they are a perpetual source of cash. Of course, many hybrids offer value for the money—increased disease resistance, higher productivity, unusual form or color—but I don't plant them if there is a good, nonhybrid (open pollinated, "heirloom") alternative.[1] This is why my other feral vegetables look like their non-hybrid parents. But my lenience toward feral vegetables should never have been extended to seeds from the compost pile, where second-generation hybrids were waiting to sneak into my garden through the back door.

Grimly, I went out to the garden armed with a two-foot Collins machete. The feral squash leered at me, like the carnivorous plant in the movie *The Little Shop of Horrors*, or perhaps more like the intelligent walking plants that conquered the world with their lethal stings in John Wyndham's terrifying novel, *The Day of the Triffids*.[2] It took a half-hour of hard work before the feral vegetables were lying on our patio—fourteen squashes totaling more than sixty pounds in weight, and at least one hundred feet of vines. The stems and leaves were saved to put back on the garden as compost, but the squashes, with their treacherous seeds, went out with the trash.

The potatoes in the garden were all right; they had been dying back in late July anyway. The tomato plants had fared worse. With the squash covering gone, they looked bedraggled and resentful. The Aunt T's still had tomatoes on them, but the other varieties were pretty much a loss. It's just not safe to be an innovative gardener in these days of agricultural multinationals, I mused. But I shouldn't blame this on agribusiness. Deep down, I felt that the giant inedible squashes were sent as a punishment for my hubris in thinking that I could reinvent in a decade a system of agriculture that my ancestors took nine or ten millennia of trial and painful error to design. I already knew that animal navigation isn't simple—why should I have thought that agriculture, which blends the complexities of nature with the complexities of human history, would be any easier? Next year, I decided, I would limit my feral vegetables to the ones I could trust.

## THINKING ABOUT BREEDS AND SPECIES

Our family dog is an Andalusian bog hound, a perfect specimen of his breed. His appearance and behavior are typical of bog hounds, but for those readers not familiar with them, a description is in order. An unusually handsome animal, he stands twenty-two inches high at the shoulder, maintains a trim weight of fifty-seven pounds, and has a short, smooth coat of a rich, red-brown color and a well-proportioned head with strong jaw musculature and brown eyes. He is extremely fast, and while running in open areas makes frequent, stiff-legged bounds straight up in the air. This may be an ancient canine technique of cooperative pack hunting, probably serving both to stir up prey and to provide a view over tall grass. His name is J. D., given by my daughter Kate because the way he moves his hindquarters when he walks reminded her of a character of that name played by Brad Pitt in the movie *Thelma and Louise*.

J. D. is exceptionally affectionate. He has never been known to growl or snap at anyone, although he once took exception to a recording of Paul Robeson singing "Old Man River." When permitted, he greets people by standing on his hind legs with his feet on their chest (or shoulders, if they are short) and gazing directly into their eyes. He rarely shows affection by licking; as a special mark of favor he may take a hand or arm into his mouth and chew gently.

His distaste for swimming—as opposed to walking in muck—is characteristic of bog hounds, especially Andalusians. We tested this behavior during a trip along the headwaters of the Passaic River, in northwest New Jersey. Coming to a deep pool, my wife, Joan, J. D.'s favorite, jumped in and pretended to be drowning. J. D. watched from the bank in great agitation as she splashed and screamed, but he made it quite plain that if anyone were to jump in and save her, it would have to be me, or perhaps another dog.

It is hard to judge J. D.'s intelligence. The famous bog hound "seven-second delay" keeps him from responding promptly to difficult commands such as "sit down," but does not seem to slow his reactions to squirrels or food. Some experts believe that the delay is an evolved adaptation to travel in bogs, where a hasty misstep can plunge an animal into treacherous quicksand.

Apart from this slight peculiarity—not really a handicap—J. D. is an ideal dog, gentle with children and friendly to adults, yet an excellent watchdog with a deep, terrifying bark; not picky about his food; not at all neurotic; equally happy on a ten-mile hike or lying on a sofa. As the vet, on one of J. D.'s rare visits, described him, he is "a great, low-maintenance dog."

After this description, some of you may be wondering where you can get an Andalusian bog hound. But I'm afraid you can't. My daughter Kate picked J. D. up while driving in the mountains of New Mexico, at a house with a sign in front that said "Puppies." (It was time, she told us later by phone from somewhere in Kentucky or Tennessee, that her brothers Sam and Jon, then seven and ten, had a dog.) The Navajo family that was giving away the pups hadn't been sure of their parentage, except that the mother was small. The puppies had purple spots on their tongues, indicating Asiatic ancestry, like a chow. There might be some wolf in the puppies, they said, but this was probably romantic invention to make them seem more desirable than ordinary dogs. Since then, we have been told, by people who were fully satisfied with their own expertise, that J. D. is, besides chow, partly Akita, partly Alsatian, and partly several other breeds whose names I have carelessly forgotten.

Given the genetic uncertainty that obscures his background, why, therefore, do I call him an Andalusian bog hound? Well, it seemed the right thing to do. You see, J. D. is so unusually good-looking that people stop me on the street and ask, "What kind of dog is that?" I used to answer, "A mutt," but gave it up because I saw their obvious disappointment. I felt obliged to supply the name of a breed. But what breed? I couldn't call him a flat-coated retriever or an Australian sheepdog because sooner or later I was bound to run into someone who actually knew what these sorts of dogs look like. Much better to find a breed that no one has ever seen; after considerable thought, I invented the Andalusian bog hound. It has been gratifying to note how much this innocent deception pleases most questioners. "Andalusian" conveys the idea of foreign nobility, while "bog hound" turns J. D. into a specialist, a distinction to which most Americans accord instant respect. True, a few cynics remain who are suspicious of J. D.'s synthetic pedigree. To these I respond by hinting at his wolf ancestry; that helps to divert them.

This state of affairs continued for several years, and there would have been no need to write this account were it not for a discussion I had one Saturday in May with an acquaintance of mine. Paul also walks a dog, and people sometimes ask what breed it is. "That question makes me furious," he said. "What difference does it make what breed it is? If they asked 'Is he friendly?' or 'Does he bite?' I could understand it. Those are questions about what my dog is really like. But asking about his breed is like asking about somebody's ethnic background. Who cares what his parents were? This kind of racial stereotyping is what has caused all the murder, torture, and destruction in the Balkans and in Africa."

I was taken aback. I had never considered that promoting J. D. as a purebred dog might be a racist act. Then I remembered another incident that had raised the issue of stereotyping in a slightly different way. It had occurred the month before, while I was giving a lecture at Yale. During the question and answer period after the talk, the subject of the introduction of exotic species came up. I mentioned the damage caused by some nonnative species such as water hyacinths, Asian tiger mosquitoes, Argentine (fire) ants, kudzu vine, and chestnut blight. Up went a hand in the audience. Was I aware, the fellow asked, of recent papers pointing out that the fear of alien species is no different from the racist hatred of newly arrived immigrant peoples? After all, he said, species have always moved from place to place. It's a natural process. Why single out the recent arrivals for condemnation? I tried to explain that introduced species, which find few predators, parasites, or diseases to keep them in check in their new environment, sometimes run amok. Water hyacinths clog waterways, tiger mosquitoes spread viral diseases, and so on. I don't think he was convinced.

Paul was outraged by the idea of stereotyping dogs by breed, and the fellow at Yale objected strongly to classifying certain plants and animals as exotic species. I asked myself, is there something wrong with using the categories of breed, variety, and species when we talk about living organisms? People love to classify things; it seems an innate part of human nature and often helps us make sense of our world. But perhaps it is a habit we should try to overcome in this new and hopeful age, when Global and One, rather than Local and Many, is the slogan, and genes can be moved, irrespective of the species differences, from cows to yeast.

Let me give this new approach a try. I'll start with dogs. Forget the breeds, a dog is a dog. No, that's not right either; I've only made the stereotype more damaging by extending it to the whole species. Better to say that each dog is its own dog. Of course, I'm still using the stereotypical d-word, but let that pass—is the statement true? Yes. J. D., like any other mutt, is unique; just as each Labrador retriever and each Siberian husky is unique. Is the statement useful? Not very. When selecting a dog to be a companion for a three-year-old child, for example, one ought to know that many but not all Siberian huskies actively resent being chased, poked, pulled, pushed, and kicked, while most but not all Labrador retrievers regard it as welcome attention. Similarly, it is sometimes useful to know that Alsatians and a few other breeds are prone to hip dysplasia; dachshunds often have back trouble; dalmatians, unlike other dogs and most mammals—but like apes and humans—excrete uric acid in their urine; and that beagles, although intellectually challenged in a conventional sense, are exceptionally good at hunting rabbits. Breed stereotypes have their value.

Despite the importance of understanding breed characteristics, it is still foolish to assume that all dogs of a particular breed are alike. A single litter of puppies may contain individuals that are timid, extroverted, loving, or unfriendly. Dobermans, famous as vicious attack dogs, often are trained as gentle companions, used to cheer up patients in nursing homes and hospitals. The interaction between the breed or species and the individual is complex. (More complexity! First animal navigation, then agriculture, and now breeds and species.) The Dutch naturalist Niko Tinbergen, who won the Nobel Prize for his studies of animal behavior, helped us understand this subtle interaction. Although he eventually learned how to tell apart individual gulls, a remarkable achievement (unless you're a gull), it was his discovery of special signaling behaviors common to *all* herring gulls that earned him his prize. Tinbergen showed us that the behavior of the individual is often comprehensible only in the context of the group.[1]

There is a caveat, of course. In *Life is a Miracle*, Wendell Berry reminds us that we always need to consider "the distinction between reduction and the thing reduced."

> To be competent a farmer must know the nature of species and breeds of animals. But the better the farmer, the more aware he or she is of the animal's individuality. "Every one is different," you hear the good stockmen say. "No two are alike." The ideal of livestock breeding over the centuries has not been to produce clones. Recognition of "type" is certainly important. But paramount is the ability to recognize the outstanding individual.[2]

Given this caveat, it is still clear that breeds and species are necessary categories. But what about the broader grouping of exotic species? True, we cannot look at a particular introduced species and tell in advance whether it will run wild in its new habitat and whether it will cause damage if it does. The gingko tree, the beautiful species introduced from China, which I planted in front of my house, does not escape from cultivation (which is one reason why I planted it). The starling, much maligned by my friend Claude, was first introduced from Europe into New York City in 1890, and now occupies much of the continent. Although people find starlings noisy, and they are aggressive toward bluebirds and some other native species, they do consume large quantities of insect pests such as cutworms, grasshoppers, and weevils. Despite these kinds of examples, however, there are more than enough cases in which exotic species have been extremely harmful to justify using the negative stereotype—with caution.[3]

The fear of characterizing breeds or species is, I suppose, a side effect of the new and long overdue political awareness of the dangers of stereotyping people according to race, religion, or ethnicity. The first of these supposed analogies, between breeds of dogs and races of people, makes no sense. The reason for this is simple. Dogs have been single-mindedly bred for specific, often bizarre traits and behaviors; people have not. Different human groups do have identifiable physical differences, such as skin color, stature, and susceptibility to certain diseases. But traits such as intelligence, character, aesthetic sensitivity, and spirituality have not been similarly sorted out by race—racists to the contrary. Intelligence and strength of character, for example, are equally valuable to people living in the Arctic, in central Africa, or in northern Europe; these qualities have moved with people wherever they went. We expect and find that they have a similar occurrence among all peoples. The other analogy, between stereotyping alien species and stigmatizing human races, is equally far-fetched. While pejorative generalizations about human races are demonstrably untrue, it is a simple matter to show that gypsy moths, kudzu vines, and Argentine ants are destructive precisely because they are alien species in a new environment.

When political correctness calls attention to the plight of oppressed peoples, it serves a good purpose. For example, the outcry against the "racial profiling" that afflicts African Americans and Hispanics in the United States is entirely proper. But when political correctness is applied to our way of characterizing different kinds of animals and plants, it can endanger the legitimate uses to which it is put.[4]

So please don't call me a racist or speciesist if you hear me complaining about the alien Asian woolly adelgid insects that are killing New Jersey's hemlocks, or the introduced zebra mussels from the Caspian Sea, which are causing so much ecological and economic damage in the waterways of eastern North America. And if you run into me while I am walking my dog, I will be happy, if asked, to express my opinions about the relative merits of beagles, huskies, Labrador retrievers, and Andalusian bog hounds.

Postscript: It is now 2008. J. D. is fifteen and one-half years old (at least eighty-five to ninety in human years, a chart in my vet's office states). He no longer runs, partly because of his age and arthritis, and partly because of an infection that damaged the ankle joint of his left hind leg, leaving him with a limp; and he sleeps more than he used to. But he still finds the book-bag that we surreptitiously leave in various rooms of the house and carefully unzips it and extracts the pieces of bread from containers that we have hidden inside. More important, we have discovered that J. D. may well belong to a real breed recognized by the American Rare Breed Association and the United Kennel Club—he looks and behaves like a Carolina dog. Carolina dogs, or American dingoes, were described in the 1970s by the biologist I. Lehr Brisbin, Jr., at the Savannah River Ecology Laboratory in South Carolina. He first found them in the pine woods and cypress swamps of the southeastern part of the state, although they occur in other places. Brisbin thinks that they are like the original dogs brought from Asia by the ancestors of the American Indians. I guess that makes J. D. exotic, although Carolina dogs are hardly numerous enough to characterize them as invasive. The ARBA and UKC classify them in the "primitive" or "pariah" group of breeds, along with Australian dingoes and Canaan dogs from Israel; and descendants of Brisbin's Carolina dogs are now available from a few breeders. So maybe you can get an Andalusian bog hound after all.

---

## STRANGERS IN OUR OWN LAND

Something odd is happening in the suburbs. I've been vaguely aware of it for a while, but it first became clear to me during a conversation with my colleague Jim Applegate, six or seven years ago. We taught graduate field ecology together and were driving a university van full of students back from an October field trip. It was twilight, and at one point we passed a small cluster of deer grazing on a suburban lawn in a development of middle-income houses. Jim, an avid outdoorsman who feeds his family much of the time on the vegetables he grows, the fish he catches, and the game he hunts, glanced at them in passing.

"This bow season I haven't been able to get one deer in the woods I usually hunt. But look at them, they're all over the suburbs where they're safe. Nobody will do anything to them here, and they know it."

Deer used to give people and their houses a fairly wide berth, but now they are eating the flowers, shrubs, and vegetables in suburban gardens, even coming up onto porches to nibble the plants in flowerpots and window boxes. The number of deer hit by cars in New Jersey has skyrocketed to many more than twenty-five thousand per year; dwellers in the suburban fringe who have not run into a deer usually know somebody who has.

Why this dramatic change in the behavior of an animal whose very name evokes an image of shyness? Previously, I just assumed that as suburban development cut into forest and farmland it brought animals more frequently into the places where people live, where food is abundant and hunting is difficult or impossible. But the more I thought about it the more I realized that there is something else going on as well; and it's affecting other creatures besides deer. A number of exceptionally adaptable species—flexible generalists—are changing their behavior to take advantage of a major shift in the behavior of people.

We have built in our suburbs an elaborate new kind of habitat, pleasing to us and to wildlife alike. It is complete with food sources such as bird feeders, succulent ornamental plantings, and garbage cans; and it has open spaces, clumps of trees and shrubs, and various kinds of shelter under and around houses—the whole area interspersed with stream corridors, forested strips, and vegetated rights-of-way. One might think we would spend every possible moment outside enjoying the paradise that we have made for ourselves at such great expense. Not so. We have gone indoors to sit in front of our television sets and computer monitors, surrendering the outdoors to the deer, geese, turkeys, crows, coyotes, foxes, bears, and others.

When I mentioned this idea to Jim, he agreed and gave me two examples to support it. One recent winter, he was driving a class along a street with single-family houses bordering on a sixty-acre nature preserve belonging to Rutgers University. It was noon of a glorious sunny day after a fresh snowfall. There were deer tracks everywhere around the houses. But not a single human footprint could be seen. In a similar vein, twenty years ago the farm fields across the road from Jim's and his wife Carol's home on the outskirts of Cranbury, New Jersey, were carved up into large lots, which were then sold and filled with a flock of new houses. Jim and Carol assumed that the extensive and accessible woods around their home would soon be overrun by the noisy children of their new neighbors. Twenty years later, they have yet to see a single child in their woods. The children are all indoors, playing electronic games or text-messaging with their friends. Biologists Oliver Pergams and Patricia Zaradic have pointed out that "after 50 years of steady increase, per capita visits to US national parks have declined since 1988. . . . We may be seeing evidence of a fundamental shift away from people's appreciation of nature (biophilia, Wilson 1984) to 'videophilia,' which we here define as 'the new human tendency to focus on sedentary activities involving electronic media.'"[1]

Richard Louv has pointed out in his book *Last Child in the Woods* that "our society is teaching young people to avoid direct experience in nature." This is, he notes, "the first generation to be raised without meaningful contact with the natural world."[2]

Some animals have been quick to exploit the new opportunities created by freedom not only from hunting but by the strange reluctance of humans to spend time outside their houses. Take wild turkeys, for example. Wild turkeys had been hunted to extinction in New Jersey by the start of the twentieth century. In the 1970s a concerted effort was made to reintroduce turkeys to the hilly, northwestern forests of the state. A few wild birds trapped in New Hampshire were released in suitable habitat. At that time, wildlife biologists believed that turkeys—wary residents of mature hardwood forests—were unlikely to survive and breed outside three or possibly four extensively wooded northwestern counties. Now, there are wild turkeys in at least twenty of New Jersey's twenty-one counties, including urban areas across the river from New York City. They have even been seen in Manhattan. These wild turkeys love the suburbs, where they move in along wooded corridors and shrubby rights-of-way, graze along busy highways, scratch around in yards, eat at bird feeders, and reproduce prolifically.

Despite their relatively tame behavior in the suburbs, when "nuisance" turkeys are trapped and released in forested areas, they fly off instantly, dropping their suburban nonchalance. This exceptional adaptability must have much to do with their current success.

Even crows, fabled for their cautious behavior, have become increasingly common in the past few years in the densely settled town where I live (although the mortality caused by West Nile disease is having an effect on their numbers). If I am walking my dog, they wait until we are fifteen yards away, or less, before flapping ponderously off a short distance to watch us while we pass. Their amused contempt for a predator on a leash, and for the helpless human leading it, is almost palpable.

The ability of some animals to make the most of the new, housebound habits of suburban people (or their harmlessness when outside) is beautifully demonstrated by the coyote. In New Jersey, when you mention coyotes most folks think of New Mexico or Arizona. Few are aware that we have thousands of coyotes—they are found in every county, where they live happily on cats, dogs, an occasional fawn, and an inexhaustible supply of rodents. In this part of the country, the coyote is more properly called the "eastern canid": on its migration from the west via Canada, it seems to have picked up some wolf genes along the way (and possibly a trace of dog). This isn't surprising because the northeastern Canadian wolves are more closely related to coyotes than to the gray wolves of Alaska, western Canada, and the northern Rockies of the United States. The eastern canid is bigger than the western coyote but in other respects is like the coyote of legend, except that there are few coyote stories here—we are too busy in virtual chat rooms or watching reality TV to note their presence.

We are similarly ignorant of our suburban foxes, except for those who make a living begging for handouts from the weekend beachgoers in some of our seaside parks. But my daughter Kate's husband, Dan, an accomplished tracker and wilderness awareness educator, finds active fox dens everywhere in the suburbs, because he looks for them. He found one, for example, in a town on the New Hampshire coast, located in a wooded strip between a McDonald's and an auto body shop on one side and a rehabilitation center on the other. Kate and Dan observed the den frequently and never saw a human footprint near it, although the den entrance and large mound of bare sand around it were clearly visible from a well-traveled path only a stone's throw away.

Easily the most dramatic suburban invasion of wildlife has been by the black bear. It is particularly striking here in New Jersey, the most densely populated state (1134/square mile). In the mid-1900s, there were fewer than one hundred bears in New Jersey, although the bears' ability to swim across the Delaware River to and from Pennsylvania makes a precise census difficult. Now the state has approximately two to three thousand of them—they have been sighted in every county. Bears are more memorable in their contacts with humans than are other suburban animals, so there is a growing folklore about them, much of it well-documented.

When people start to feed wild animals in the suburbs, trouble ensues. Bears, perhaps because they seem benign and cuddly to those whose experience of nature is provided mostly by television and Walt Disney, are often tossed bread and other inappropriate delicacies; moreover, they find it easy to open most garbage cans, which must seem to them like direct feeding. This partly explains why the number of reported instances of bears entering houses in New Jersey averaged between twenty-five and fifty annually between 2000 and 2006. This is an order of magnitude more than occurred a few decades earlier. They open kitchen cabinets and refrigerators in search of food: one startled woman returned home in the middle of the day to find a large bear in her kitchen feeding herself and two cubs on frozen chickens—if they had been eating porridge, it would have been the Goldilocks story in reverse. The bears exited through a window. No humans have been seriously injured by bears in New Jersey in recent years—nevertheless I suppose it is just a matter of time until somebody gets badly hurt or killed. But sentiments in the state seem to be decidedly pro-bear; a proposed hunting season for bears to reduce the population was cancelled in 2006 by the commissioner of environmental protection after the governor received a flood of e-mails protesting the idea.

A parallel situation exists with alligators in the American southeast. Despite their former abundance, early reports of wild alligators attacking humans were extremely rare. Alligators viewed people with massive indifference. We can glean this information from a number of sources. One of the best is *The Alligator's Life History*, by E. A. McIlhenny, of the McIlhenny's Tabasco Sauce family. In this first important book on alligators, privately published, McIlhenny gave a remarkably accurate account of alligator biology, including anatomy, feeding habits, vocalizations, nesting, and the first description of strong maternal behavior in this, or any, reptile. Initially discounted by academics as the work of an amateur, his findings are now recognized as correct. McIlhenny grew up in the latter part of the nineteenth century on Avery Island, Louisiana, a six-thousand-acre tract that his family owned. He described his early contact with alligators as follows:

> Among the earliest remembrance of my childhood is running down with my brothers and cousins and other small boys in the warm summer afternoons to the boathouse to swim. . . . The bayou is about one hundred and fifty feet wide at this point and about ten feet in depth. . . . In these days the alligators in the streams about the place were more than numerous, and of course, boy-like we always took great pleasure and not a little excitement in seeing how many 'gators we could call around us during our

swim. We would attract them by imitating the barks and cries of dogs and by making loud popping noises with our lips . . . and they would come swimming towards us from all directions. We had no fear of them and would swim around the big fellows, dive under them and sometimes treat them with great disrespect by bringing handfuls of mud from the bottom and "chunking" it in their eyes.

In 1879, the largest of these alligators, nicknamed "Monsurat," one the boys "never took liberties with," was shot and killed by young McIlhenny's tutor from Harvard. It measured eighteen feet, three inches, and took four mules to drag to the house.[3]

Hide hunting changed the relationship between alligators and people. Intensive hide hunting is now largely a thing of the past, but tourists and others sometimes feed the reptiles, which is almost as bad. Alligators that have had the experience of being fed can become dangerous and unpredictable; there have been a number of fatal attacks in Florida. People whose experience of nature is entirely virtual, mostly gained from television, do not always know how to behave appropriately when faced with real animals in their environment.

The resurgence of wildlife in the semi-deserted environs of the "Glen Manors" and "Hillside Estates" tells us something about the new lifestyle of Americans. For the habit of staying indoors is not just a suburban phenomenon, although this is where it is most apparent. The number of hunters is decreasing steadily all over the United States (but the proportion of hunters who are women appears to be rising). Hikers, bird-watchers, and amateur botanists are numerous and vocal in defense of nature; but they remain a small percentage of the population. On any beautiful summer weekend in the forested parts of the New York–New Jersey region, there are few people on our magnificent hiking trails once you get a quarter of a mile from the parking lot.

In his environmental history of the twentieth century, *Something New under the Sun*, historian J. R. McNeill describes two mutually exclusive life strategies that animals, including humans, may follow.[4] One strategy, the rat strategy—which we could also call the deer, turkey, bear strategy— is to remain adaptable and flexible, changing behavior and outlook to suit a changing environment. Until recently, people have been accomplished rat-type strategists; some individuals and societies still are. For these, life is inseparable from the outdoors and the nature around them. The other strategy—of inflexible specialization—is to make a supreme adjustment to one particular favorable and constant environment, to give up flexibility so that this environment can be completely exploited. Pandas do this: they eat only the succulent tips of young bamboo, a dietary rigidity made possible by the fact that bamboo is widespread and usually abundant, although bamboo growth periodically fails and panda numbers crash accordingly. Suburbanites and many other residents of the industrial world have done this too: we have staked our future on the assumption that electricity and carbon-based fuels, which support our indoor existence, will always be plentiful and affordable. The panda's gamble appears far safer than ours.

To specialize in a life revolving around the electric outlet, the light switch, the Internet, and the gasoline pump is frightening, because the cheap and reliable supply of these modern, energy-consuming "necessities" is so potentially fragile. As specialists, we have cut ourselves off from the rest of nature, thinking one of the supremely foolish thoughts in the history of our species, that we no longer depend on or need to understand the natural world. So the deer, turkeys, crows, Canada geese, coyotes, foxes, and bears in the U.S. suburbs are not really a confused remnant of wild nature displaced by development, which our misplaced sense of superiority would have us believe. Rather, they are the vanguard of a group of canny generalists—from vertebrates to viruses—poised to take over, at least temporarily, an environment that we seem to have abandoned.

But I don't want to end on a note that is overly bleak. Nature awareness is alive and well in at least one segment of our society: bird-watchers, weekend botanists, mushroom hunters, and other amateur naturalists are thriving, if the profusion of modern field guides available to them and the sales of binoculars and spotting scopes are any indication. Not a large percentage of the population compared with the number of fans of football, NASCAR racing, or reality TV, they nevertheless command considerable economic clout. Moreover, they include enough young people to give me hope that our contact with nature cannot be completely suppressed—there will always be people who can guide us back to an understanding and appreciation of the natural environment in which we live and the nonhuman creatures who share it with us.[5]

---

## TEACHING FIELD ECOLOGY

The first thing I did when I arrived at Rutgers in the late summer of 1974 was to plan the courses I would teach. My principal fall course was to be based on one that I had helped teach for a few years at Barnard College: The Natural History of the New York Area. At Barnard, I had learned the subject by accompanying far more experienced colleagues— Tony Warburton, an evolutionary biologist, and Patricia Dudley, an ecologist—on their field trips. Now, in New Brunswick, I had a new teaching partner, Jim Applegate, a wildlife biologist, but I didn't anticipate any changes.

Jim listened to my plans for the course with gratifying attention and enthusiasm. He had only a few questions. "What are we going to call the course?"

"'The Natural History of the New York Area,'" I answered, "or maybe 'The Natural History of New Jersey.' That's what it is, isn't it?"

"Sure. But we already have our course in General Ecology, which you are going to run. That's mostly theoretical, indoor classroom learning. Why not call the new course 'Field Ecology' and design it to let students who have had General Ecology apply their knowledge to the real world? In other words, we want to teach them more than descriptive natural history—they should understand the ecological and human processes that make each place what it is."

This meant a pretty complete rethinking of the course, which I hadn't expected to do, but I grudgingly agreed. Thus began what has become the most remarkable experience of my teaching career.

For the first three or four years, we taught together: two different sections a week, each with the two of us and fourteen students crammed into a fifteen-passenger van for field trips that lasted from 1:00 to 6:30 P.M. From the start we decided that there would be almost no classroom teaching, just field trips, regardless of weather. And so we have witnessed the majestic silence of a white cedar swamp in the October sunshine, have walked the springy, low-tide-bared *Spartina* salt marsh in torrential rain, and have given final exams on abandoned landfills during snowstorms. (The best technique for taking an exam in the snow or rain is to put a clipboard and paper in a large, clear plastic bag and write inside the bag with a medium pencil.)

Jim and I worked well as a team, the born-and-bred outdoorsman and the suburban-grown theoretician. He could show the students exactly why the Pine Barrens vegetation and accumulating litter layer in the new paradise-in-the-pines luxury housing development meant that the whole complex was constantly at risk of being incinerated in a forest fire; I could explain Lewis Mumford's insights about places like that and how such developments failed even when they did not burn. As we walked through the Hutcheson Forest, Rutgers's sixty-acre, old-growth wilderness ten miles west of the college, Jim would point out all the exotic plants—Norway maple, Japanese honeysuckle, and *Ailanthus*—that were displacing the native vegetation, and I would discuss the various explanations of why exotic species are able to invade and alter the character of a supposedly protected habitat.

Almost of its own momentum, a theme developed that included and greatly transcended my first idea of a regional natural history course. In Field Ecology we expect our students to become able to enter any landscape—familiar or strange—and, by seeing its plant and animal life, its landforms, and its human artifacts, to describe its history, its present condition and use, and its fate in the foreseeable future.

We start small. For example, on the first day of the course in 1992, we wandered around the campus, introducing our methodology to the students, all of whom already had taken General Ecology. At one point, I stopped at a four-foot, bushy Colorado spruce that Jim had previously pointed out to me, and asked them to tell me what it told them. They crowded around the little tree, examining and poking.

"It has multiple trunks," said one observant forestry major, "so it must have resprouted after being burned, or cut, or grazed."

"Not burned or grazed," said another student. "Not here in the middle of campus."

"Right," I agreed. "So tell me in what week of what month of what year it was cut, and at what time of day."

They stared at me blankly. "You've got to be kidding," an exasperated senior muttered. A minute or two passed, then a few smiles appeared simultaneously.

"The week, month, and time of day are easy—it was cut the second or third week of December, after dark," one student offered.

"That's right. And the year?" They gingerly pulled the sharp-needled stems apart, counting the annual growth intervals back to the point where the trunks diverged.

"1988," they announced after a brief conference.

During this time one student had stayed aloof, a funny expression on her face. I asked her what was the matter. She hesitated, then answered: "I know the fellows who cut it—they were in my freshman dorm. It was a couple of weeks before Christmas, in 1988. At night, so nobody would see. I helped them decorate it." She looked at her classmates in amazement. "I can't believe that you guys figured all that out just by looking at the tree."

From single trees we progressed to whole ecosystems. The presence or absence of wild grape and bittersweet vines, flat versus hummocky topography, invasion or lack of invasion by exotic species such as winged euonymus and *Paulownia*, a strong or indistinct deer browse line, regeneration or no regeneration of tree seedlings, pin oaks versus chestnut oaks, Japanese barberry versus maple-leaved viburnum, white snakeroot versus highbush blueberry and sphagnum moss, caddis fly and mayfly larvae versus *Tubifex* worms and *Physa* snails . . . these were the elements of which stories were made, histories were deciphered.

After a few years, Jim and I stopped teaching Field Ecology together. Two professors for fifteen students twice a week was too great a luxury for a modern research university. I was suddenly on my own, and very nervous. Fortunately, the students were well trained in the subjects I had never had in college: dendrology (the study of trees), geology, soils, wildlife ecology. "Why is this black locust here?" I would ask, pointing to a handsome tree with a strongly patterned bark that identified it to me.

"Er, I think that's a sassafras, sir," said one of the star dendrology students, politely. I didn't argue; after all, she was helping to teach the dendrology course. Besides, I rationalized to myself, it's good for them to see that the professor is fallible. And there was my wonderful memory of Professor Nash, one of my chemistry teachers at Harvard, whose lecture demonstrations were often carefully planned to turn the wrong color, or even blow up, because he felt that nobody learned anything from demonstrations that worked properly. Of course the analogy was backward: I was the one who was learning from the mistakes.

Meanwhile, in the other section, Jim dropped the urban/suburban field trip because he had been uncomfortable with it and added an exercise that I haven't yet had the guts to try. It's called "Stump the Professor." One year, for example, two of his best students brought Jim and the rest of the class to a completely unfamiliar forested area that they had secretly researched in advance—the way Jim and I scout out our trips ahead of time. It took Jim about half an hour to figure out that it had been a dairy farm until the 1940s. Nobody else got it, although the better students came close.

In recent times I have noticed a difference in the training of the students who come on my field trips. They are just as bright as before, but many of them seem to have had little direct experience of the natural world while in college. Here's an illustration. We have a wetland field trip, designed by Joan, that begins with a visit to the floodplain of the Raritan River. I show them that in the floodplain the riverbank is highest right next to the river, because during floods the waters slow down as they pour over the bank, dropping their coarsest and heaviest suspended gravels and sands first, thus building a natural levee along the river. As the floodwaters move farther from the river, they slow down even more and deposit a thinner layer of lighter particles, fine silt and clay. The two zones of the floodplain have very different vegetation; the higher, better-drained zone next to the river has many species of trees and shrubs, while the lower, wetter zone farther from the river, which is more frequently flooded by rapidly running water (when waters pour through gaps in the levee), has fewer species of trees and almost no shrubs. Recently, at Joan's suggestion, I started bringing along a soil auger so that we could get actual soil samples rather than take the theory for granted.

"Most of you have had Soils, right?" The students nodded. "Then you've used soil augers." They shook their heads. That wasn't part of the course. Not their fault, I realized, but it got me upset anyway. Going out to look at the natural world and to learn from it is deemed too expensive by cash-hungry universities (although a miniscule fraction of the cost of equipping a gene-splicing lab), and why spend resources on an unglamorous subject like soils? Needless to say, the student products of this new, virtual education can be woefully helpless and infuriatingly uncreative when let loose in a real environment. And the problem is compounded by the fact that many of them spent little time outdoors before college—they were too busy text-messaging, surfing the Net, and playing electronic games.

But the field trips are still great fun. Also, New Jersey may be the best place in the United States in which to teach field ecology. Nowhere else is such a bewildering diversity of mineral types, geological formations, plants, and animals, and such a continuum of rural-to-urban human settlements crowded into one small area. Within a fifty-mile radius of Rutgers, we can examine a trout stream strewn with boulders in a hemlock-lined gorge. We can visit a holly forest dotted with cactus on a narrow barrier beach. We can stand among the shoulder-high but fully mature pitch pines and blackjack oaks in one of the mysterious patches of dwarf forest in the Pinelands. We can look at the amazing differences in vegetation between neighborhoods that were settled by immigrants in the 1930s and those built by developers in the 1990s. Or, in this state whose population density is greater than Japan's, we can explore sparsely settled semiwilderness areas that contained more towns and people 150 years ago than they do today.

Finally, for me at least if not for the students, there is the exhilarating, terrifying, enlightening sense of the power of the passage of time that comes with paying close attention to the same places year after year. Imagine what it is like to show the class lawns, flood retention basins, and town houses where prior classes have watched potatoes being harvested, and to explain why the only crops they will see on their agricultural field trip are corn, soybeans, and Christmas trees, while their predecessors could also visit orchards and multicrop truck farms. Then erase that image and picture yourself standing with the class on the eroding secondary dune of an ocean beach, holding the faded map of the self-guiding nature trail that you walked with other students twenty-five years ago. Now imagine yourself telling them that stops 1 through

25 on the trail—the ghost red cedar forest, the gnarled black cherry, the giant poison ivy vine, each with its numbered signpost corresponding to the numbers on the map—are, as you speak, one hundred yards out at sea, like the drowned villages of the Lenape, deep under the waves. Do you feel small? Do you feel transient? Do you feel old?

Here are the two great kinds of change we encounter in life, cultural and ecological, bobbing and weaving together, forever repelling and embracing, inexorably meeting at the myriad points of our own mortality. Field Ecology has let me see this ghostly process in a sharper light than is given to that sad majority that lives—for a while—only in the dim and static world of its own unnatural contrivance. But though it saddens and frightens me to have relentless time made so vivid and so personal, I can yet enjoy watching the stately and immensely complex dance of culture and nature, and can understand a little better the society and place in which I live.[1]

---

## THE UBIQUITOUS RIGHT-OF-WAY

On all but one of the field trips in my Field Ecology course, I take my students to the sorts of places that most of them have been to before: the beach, the pinelands, the Highlands forest, farms, old fields, streams, salt marshes, suburbs. But on the third trip, after the ones to the campus and to the experimental plots at Hutcheson Forest, we go to a place that is, for all its superficial familiarity, altogether different and exotic. This trip is to America's deserted empire, what the person who knew it best, ecologist Frank Egler, once described as the Rightofway Domain.[1]

Right-of-way land comprises at least fifty to seventy-five million acres in the United States, an area larger than New England. It is disposed as long strips of property along railroad tracks, roads, and canals; under power lines; and above buried pipelines. Many of these rights-of-way, even those in heavily populated areas, are scarcely ever visited by people—they are cut off from human presence by fences, no-trespassing signs, patrolling police, dense vegetation, and a paucity of reasons to set foot in them.

True, some abandoned rights-of-way have been put to use. The preserved towpath and adjacent land along the old Delaware-Raritan Canal and its feeder canal, which wind for more than fifty miles through central New Jersey, have become a very long, very narrow, very popular state park. Indeed, this is the park that has helped protect the forest along the Millstone River, which I described earlier. Hunters love the rights-of-way under power lines, which attract deer and small game. And disused railroad lines have been turned into foot and bike trails in several parts of the country. But many rights-of-way, totaling a huge amount of land, go for months or years without feeling a human step or hearing a human voice. These places are themselves neither urban nor suburban nor rural; neither settled nor wilderness. They are a quintessential part of what author James Howard Kunstler has called "the geography of nowhere."[2]

In my right-of-way trip we start with a boggy strip of land above a transcontinental gas pipeline. The students examine the vegetation and deduce that it is maintained by mowing, in the fall every two or three years. From the pipeline, we go to a railroad line to observe the effects of the powerful herbicides that are sprayed on the plants growing along the tracks by moving tank cars. We used to spend some time walking beside the mainline tracks of the Northeast Corridor, which runs from Washington to Boston. Then one day we were stopped by a large man with a dog, a gun, a rattlesnake belt buckle, and an Amtrak badge. The dog was a well-developed German shepherd, not wagging its tail. The man seemed interested in what our class was doing; he asked questions. Then he told us to go away—and we went. Since then, we have taken a quick, furtive look at the tracks from a protected vantage point behind a factory.

The next stop in our trip is a fascinating brushy strip under a high-voltage power line. The towers that support the power cables dwarf the red maples, sweet gums, gray birches, and pin oaks that have grown up in this damp right-of-way. It takes the students only a few minutes to figure out that the power company fells the weedy trees under the line approximately every ten to twenty years, using both bulldozers and chain saws. I don't like to spend too much time there because of the extra-low frequency electromagnetic fields that the cables are generating. I use a meter to show that the field strength on the ground under the power lines is as high as 60 milligauss. Maybe the British National Radiological Protection Board is wrong and electromagnetic fields above four milligauss do not double the incidence of leukemia in children, but why take a chance?[3] Some of the students wonder whether the people living in the apartments next to the power line suffer any ill effects. I tell them I doubt it—the fields quickly diminish with distance, as I show them with my milligauss meter. But if their children play under the power line, they may not be so secure.

Naturally, I save the most exciting stop for last. Driving the van for a short distance down Interstate 287, westbound, I tell the students what I know about the planting and care of the right-of-way land that runs along the highway margins. In the first year we did this field trip, Jim and I experimented with pulling over and letting the students get out and look. But the noise of the cars and eighteen-wheelers made it hard for the students to hear us, and the state police were not happy with our stopping there, so we had to find a better way of teaching about the interstate's right-of-way.

Now, after a mile or two on the interstate, when I get to State Route 27, I take the southbound exit around the cloverleaf, drive on Route 27 under the superhighway, pull into an adjacent chemical company's parking lot immediately on the other side of I-287, notify security that we have arrived, and park the vans. With the students following, I walk across the parking lot to the exit ramp from eastbound I-287. Waiting for a break in the exiting traffic, we cross the roadway and then cross the parallel entrance ramp leading from Route 27 onto the interstate. If you have followed me so far, you know that we are now in one of the four leaves of the cloverleaf. This is Rightofway Island, our destination.

About five acres in area, the nearly circular island has concentric bands of vegetation. Next to the entrance ramp we just crossed, which forms about three-fourths of the island's perimeter, is a narrow zone of grasses mixed with the pink-flowered, ground-hugging relative of clover called crown vetch. The vetch is probably of the variety known as "Interstate," which was developed as a low-maintenance planting along interstate highways. This zone is mowed occasionally. Next to it is another zone, maybe one hundred feet wide, with tall weeds and low-growing bushes. In the middle is a zone of trees. On the far side from where we are standing, the island is bordered for about one-fourth of its

circumference by the high, vetch-planted bank that supports the interstate, out of sight but not out of hearing, above.

We move hastily through the weedy zone, stopping only to eat a few of
the everlasting peas that have ripened in the September sun. Our goal is
the dumbbell-shaped patch of woods at the center of Rightofway Island;
I won't be able to relax until we are safely in its shade, not visible from
passing cars. The last barrier between the woods and us is a belt of multi-
flora rose, a thick tangle of intertwined canes studded with stout thorns.
But the path we made through it last year and the year before that is still
there, and we arrive at the little grove with only a few minor scratches.

The trees of the grove are nearly all sassafras and nearly all are the
same age, about thirty-five to forty-five years. They must have sprung
from the seeds in berries dropped by birds soon after the interstate, with
its cloverleaf, was built. The trees have grown thickly, and the shade has
kept the ground pretty free of vegetation, except for an occasional
creeper of Japanese honeysuckle where gaps in the canopy let in some
light. Evidently this spot never received any fill during the construction
of the interstate, so it is a bit lower than the rest of the island. We cannot
see out beyond the encircling bushes.

I find a place to sit up against a tree trunk. The students follow suit.
Soon, despite the ever-present roar of the interstate and the lesser growl
of Route 27, a novel awareness comes to most of them—I can see it in
their faces. We are *alone*, they realize, cut off from the civilization
around us.

Like Dorothy and Toto in Oz, nobody in the world we have left can
see us or hear us—nobody else from our world knows that this place
exists. There are no empty cigarette packs or beer bottles on the ground,
no footprints except ours; not one living soul has been here since my last
visit, a year ago. We are conscious of the world we have left—the noise
and the polluted air accomplish that—but, as if we had passed out of familiar time and space, our world is no longer conscious of us.

Here is the New Wilderness, isolated from surging humanity more
effectively and completely than the inner reaches of the jungles of the
River Amazon. No ecotourists will ever come here. The little grove
exerts a kind of weird fascination; it casts a spell, but not the spell of a
magical woodland or enchanted garden in a fairy tale. There is a quality
of magic in the isolation of this place, but it has an incomplete,

purposeless, unsettling feeling. This is Oz with a difference. There is no yellow brick road in sight.

Not much lives in this grove, aside from the trees. There are no lichens on the bark, no snakes on the ground, and few birds among the branches. We have not found any nests. One of the students sees a rabbit—there are probably field mice and voles out in the vetch. A raccoon or skunk able to make it alive past the encircling lines of traffic would have to run the gauntlet again every time it wanted water. Once, when we were seated among the sassafras trees, a young deer wandered by a few feet away, moaning loudly. But Rightofway Island was not always like this.

In the center of the grove are two mounds of rotting wood. Twenty years ago, they were two dead trees, standing about fifteen feet apart. Then the students could still identify them as apple trees and could understand their significance. There is meaning in two apple trees growing not far from Route 27, since colonial times a part of the main road from Philadelphia to Princeton to New Brunswick to New York. This island was a place where people used to live, in the days before the interstate, before the chemical company, before the strip malls, and before the discount box stores. Perhaps the trees were part of an orchard—more likely, they were in the yard of a house. Now, nobody will ever live here again, not even after the interstate highway, transient relic of the age of oil, has carried its last truck howling into the night. The adjacent massive bank of fill that supports the interstate roadway has made this an unsuitable site on which to build, at least until the next glacier scrapes it away. Nor can it, while the highway lasts, return to what it once was, an eastern deciduous forest with a great diversity of plants and animals. Because of the traffic pattern, one could not even put a fast food restaurant or gas station here.

I explain the mounds of decayed wood to my students and ask them to think of the tens of thousands of other pieces of land cut off from people and nature by superhighways. The ecologists William Niering and Richard Goodwin showed how, at little cost, we can help the land under power lines support huckleberry bushes and quail, but there is nothing comparable that can be done about this kind of place.[4] It is too small and too isolated by unceasing traffic; it has no water. A mere fragment, the island has no connection with any adjacent, larger woodland that could serve as an enriching and rejuvenating reservoir of plants and animals. No squirrel will ever bring an acorn here. It is in limbo, like so many of our works of progress, accessible to neither the creativity of humanity nor the creativity of the natural world. Is this a preview of remnant nature in our midst? A resident alien, cut off from its larger, life-giving self? If so, what is to become of *us*?

The sun is low in the west and it is time for us to leave. Once again we successfully cross the two lanes of opposing traffic. A truck driver honks at the female students; we have left Oz. I look back at Rightofway Island and think that of all the places we visit in Field Ecology, it is the strangest and most sad. Everyone is anxious to get back to the van. As we drive up to the gate of the lot, the security guard smiles and waves us on.

---

## A WALK IN THE WOODS

The forest was small by American standards, perhaps fifty or sixty acres, but in the rolling Devon countryside with its parceled fields and narrow, hedge-enclosed lanes, it felt appropriately spacious. I was enjoying the guided tour in one of my favorite parts of the world. The light rain and chilly July breezes felt right; the English ivy carpeting the ground seemed right; the leafy hardwoods looked right (although I didn't know the species and could easily have been fooled); and the probably medieval bank and ditch running through the woods at right angles to the path gave everything an impressive air of authenticity. The path turned. As we rounded the corner, I saw ahead the darker shade of conifers. Soon we were in the midst of a grove of youthful but already towering California-coast redwoods. A deep silence hung like a benediction over the dark wood, but it was quickly shattered. "I'd give anything to be allowed to cut them down," said our guide, Stephan, in an angry voice.

This incident passed out of my thoughts until weeks later back in New Jersey, when I was reading the chapter on conservation in Oliver Rackham's *The History of the Countryside*, an account of the origins of Britain's landscapes, flora, and fauna.[1] The contemporary British land-scape, Rackham wrote, is suffering from an acute loss of meaning—the unique messages once conveyed by many historic woodlands, witness to millennia of slow and painstaking change, have been garbled beyond recognition in five or six decades of modern planting, "restoration," development, and agriculture. The more I read in this remarkable book, surely one of the most profound and eloquent descriptions of people and nature ever written, the more I understood Stephan's feeling that his Devon woodland had been desecrated by the planting of those redwoods. I also began to understand how little I knew about the long discourse between people and trees in Britain, where the history of the relationship is probably as well documented as in any place on earth.

As a resident of eastern North America, much of which is wooded, I have known since early childhood what a forest is. It is a place where many trees grow together, a place defined by its trees. In a forest, the human presence is optional and variable. Humans can have a great effect on forests—or hardly any—but we are external to the basic idea of forest. We use the forest, manage the forest, abuse the forest, destroy the forest, protect the forest, enjoy the forest, fear the forest, study the forest, and ignore the forest; always the forest is measured against an original Forest in which people do not appear at all. (This idea of Forest ignores the major role played by American Indians in shaping the structure of North American forests for thousands of years before the Europeans arrived.)

As it turns out, the word *forest* throughout its long history has had a much weaker relationship to trees and a much stronger relationship to people than its current meaning implies. The original Latin word *foris*, from which *forest* evolved, simply meant "outside." In England, the association of forest with trees was even weaker. The word was first introduced in the eleventh century by William the Conqueror; it meant an officially designated place where deer were protected and their usage governed by special laws and regulations. A typical forest might have included villages, towns, agricultural fields, and nonwooded countryside, in addition to patches of woodland. (The twenty-first century discipline of "urban forestry" is an echo of this older definition.[2]) As Rackham points out, the designated forests in Britain were mostly in the less wooded regions, and woods were usually only a small part of them. Some of the largest forests had little woodland; for example, Dartmoor Forest was nearly all moorland and Sherwood Forest was primarily heath. Robin Hood and his merry men must have spent very little time hiding behind trees.

Most official forests were under the jurisdiction of the king; some were controlled by nobles or by senior clergy. The forestal rights included primarily the right to take deer (or honorary deer, such as pigs), the right to appoint the sizeable bureaucracy that administered the forest—wardens, seneschals, justiciars, verderers, and so on—and the right to collect fines for violations of forest regulations. A gift of game in honor of a wedding, graduation, pregnancy, or other festive occasion, or the gift of an appointment as under-forester, had great political value for a cash-poor medieval king.

The benefits the forests provided, however, included far more than the collection of game and the privilege of appointing administrators. There was also the ownership of the land, the mineral rights, rights of pasture, and, in the wooded areas, the rights to timber and to the underwood, the multiple stems that resprouted from the stumps or roots of cut trees. Rarely did all these rights reside in the same hands. For example, a nobleman might own the land, the king might control the game, timber, and administration, and the commoners often had the valuable rights to the underwood and pasture. While kings frequently quarreled with the nobles over forest privileges, they tended to respect the rights of commoners, which predated the designation of the forests. Not until the enclosures of the eighteenth and nineteenth centuries did many of these forests pass exclusively into the hands of the landowners, who often promptly cleared the wooded portions for cropland.

But many patches of woods, some large, remain in the surviving official forests and in the originally wooded parts of the country that never were in the forest system. This British woodland is not an unchanged remnant of the wildwood that covered the land six thousand years ago, when the first Neolithic settlers arrived and began, with amazing efficiency, to convert the landscape to open, agricultural fields. Although the kinds of trees once found in the wildwood still exist, the woodland itself is now something different, as much a cultural as a natural environment.

One difference is a striking change in the relative abundances of the various tree species. In the wildwood that spread over much of lowland England, for example, the dominant tree was the stately lime or pry, *Tilia cordata.* (A close relative in the northeastern and northcentral United States is *Tilia americana*, the American basswood, or linden.) We know that the lime was common from the record of preserved pollen grains. Now it is rare over much of its former range, while oak, ash, and beech, which like starlings and raccoons are comfortable around people, have increased at the lime's expense. One reason may be that the lime is particularly attractive to the herbivores that people introduce and favor—deer, rabbits, and sheep; its American cousin, the linden, is similarly sensitive to browsing. There are no doubt other reasons why the lime gets along uneasily with British civilization, but they are hard to fathom. In south Suffolk, Rackham writes, lime is still fairly abundant in a few woodlands that are known or believed to be ancient, but in places where these woods are expanding into adjacent fields, the lime is reluctant to go beyond the old woodland boundaries that were set in Neolithic times. Some flowering plants, such as the oxlip, *Primula elatior*, have the same reticent behavior.

Another critical difference between the wildwood and the woodlands of today is the practice of coppicing—cutting the underwood every few years. Coppicing has been a hallmark of British woodsmanship since the Stone Age and is one of the best examples of sustainable forestry anywhere. Underwood has had many structural uses and has been a major source of fuel. When done properly, coppicing can continue indefinitely; Rackham notes that the Bradfield Woods have been cut at least seventy times and are thriving. An ash that would normally have a lifespan of two hundred years can live and be healthy for more than a millennium if coppiced, its stump, or "stool," spreading in rings that are as much as eighteen feet across. In wooded areas that were heavily foraged by game or cattle, the valuable resprouting stems were likely to be eaten, so trees were frequently pollarded—cut at a height of eight to twelve feet above the ground. This put the new growth of stems above the reach of browsing animals. Pollards were repeatedly cut for wood, like coppice.

A woodland with a mixture of coppice and timber is not only diverse and attractive, but is also immensely durable, yielding products of great value century after century at minimal cost to the people who use it, while retaining the species of plants and animals that give it its distinctive and stable character. This is why much of the British woodland extant in 1700 would have been familiar to an Anglo-Saxon resident in the year 1000.

Since 1700, many acres of woodland have been lost, some to plantation forestry and ill-conceived restoration projects, but mostly to agriculture—first through the enclosures and recently through the rise of industrial agribusiness. The conversion of woodland to agricultural fields has been, overall, both a social and an economic failure. Because woodland usually is found in places unsuitable for agriculture, and because modern industrial, monocrop farming is vulnerable to disturbance by pests and diseases and expensive to maintain as energy costs rise, the replacement of woods by farms has yielded little or no lasting benefit to their owners.

In the remaining woodland, the art of coppicing has also fallen on hard times. At present, oil is easier (but not necessarily cheaper) than underwood to use as fuel; other materials have replaced coppiced stems; timber is available from Canada and Scandinavia; and forest foods such as hazelnuts are hardly a necessary part of the British diet. Yet coppicing has not died; in the 1990s it was revived in a conscious effort to maintain historic practices. It is a source of firewood to be used instead of oil, provides materials for special crafts, and creates animal and plant habitat. Whether coppiced woodlands, one of the oldest and most successful compromises between people and nature, will have a new role to play in the coming era of scarcities remains to be seen. It is not out of the question.

In the meantime, woodlands still grow all over Britain, and some men and women still know how to care for them. I take it as an encouraging sign that there are people who can rage against the planting of a redwood grove in a Devon wood, who can resent and resist the breach of faith and loss of meaning that mark so many of our dealings with the landscape. Long in the making and long in the proving, the ancient compact between Britons and the natural world may yet survive the uncertain centuries ahead.[3]

---

## OLD GROWTH

The commentator Howard Berkes, on National Public Radio's *All Things Considered*, said something about planting trees, and I tuned back in. When I'm driving, I have the bad habit of keeping the radio on, while mentally tuning out the commercials and the ceaseless, repetitive, talk show babble that has taken over WNYC's programming since the September 11 attack knocked out the transmitter for their classical music broadcasts. But this talk was about trees, "champion trees," and I turned up the volume. What was a "champion tree," I wondered?[1]

Champion trees, it turned out, are defined as the biggest, oldest trees of any given species. There is a Champion Tree Project whose aim is to identify more than eight hundred of these trees from every part of the country, harvest their buds, and clone them to provide champion off-spring saplings for a variety of gala tree-planting events. The Champion Tree Project was created in 1996 by David Milarch, a Michigan tree farmer, and his son Jared. In promotional literature that I obtained later, David Milarch wrote, "These old forest giants had the genes and good luck to survive recent centuries of industrialization, development and deforestation." And on NPR, Milarch explained his use of the term *champion*: "A champion, in our opinion, is the guy that's left standing after the rest have fallen. There's only one champion," he said.

The announcer went on to describe two high-profile plantings of champion trees, the planned memorial grove at the Pentagon to honor the victims of September 11 and the reforestation of the Salt Lake City area, including sixty-five champion clones, for the Olympic Winter Games of 2002. The Pentagon grove was sponsored by the National Tree Trust. Its director, retired Marine Corps general George Cates, pointed out that the champion trees have survived "the worst that Mother Nature and man can throw at them over time . . . they have seen every war, they have seen every turmoil." The example of the champion trees will provide hope to the American public, General Cates announced: the grove will say, " 'We've seen it all, we've come through it and we're going to do it again.' " The NPR report concluded; I turned off the radio and drove on in silence.

I was all in favor, I decided after a few miles, of planting trees, champions and others, around Salt Lake City to celebrate the Winter Olympics; and I think that a living grove of trees at the Pentagon is the best kind of memorial that there could possibly be to the innocent victims of the vicious attacks of September 11. Any caveats I have about champion trees do not apply to the original vision of David and Jared Milarch or to the two admirable projects I have described. Nevertheless, the quintessentially American idea that for trees (and for so many other things), the biggest = oldest = toughest = best gave me pause.

I thought back to the field trip my wife, Joan, and I took to northern Quebec's Ungava Peninsula in the summer of 2000. The trip was sponsored by the grand millennium meeting of wetland scientists from all over the world, which Joan was attending. We were flying in a Twin Otter plane, an Air Inuit charter, heading for Professor Serge Payette's remote northern field camp next to the Boniface River.

When we had taken off from the airport at Kuujjaraapik, on the eastern shore of Hudson Bay, I had been told that Air Inuit's two most senior bush pilots were at the controls. Now, as we circled the tiny landing strip carved out of the brushy vegetation on a low bluff above the Boniface River, I knew why. The distressingly short runway took a thirty degree turn in the middle, something I had never seen before and don't care to see again. But the pilot seemed satisfied; he had a brief conference with his copilot, gave the thumbs up sign, and we began our final approach.

After the pleasantly uneventful landing, we disembarked to find Serge Payette waiting for us. "Why does the runway do that?" I asked him, as we unloaded fifty-five-gallon fuel drums from the plane.

"We had to avoid a wet spot," he explained, and before I could question him further, went to help head off a drum that was rolling toward the river.

The Boniface River camp was both basic and elegant. We slept in a tent, made use of a primitive outhouse, and had no place to shower, but we were given wonderful dinners—for example, fresh-caught fish, which I think were trout, in sauce amandine, with a choice of red or white wine—cooked by a superb chef who had flown in with us on the plane. The day after our arrival, after a substantial breakfast, Serge Payette took us out to his study sites in the peatlands and wooded areas near the river.

A legendary figure, both for his vast knowledge and his charismatic presence, Payette has trained many of the best ecologists in eastern Canada. He is much of the force behind the dendrochronology lab at Quebec's Université Laval, and is the former director of the university's Centre d'Études Nordiques, with a field station located in the twin

Inuit/Cree towns of Kuujjaraapik/Whapmagoostui. His learning encompasses and integrates botany, geology, soil science, geography, climatology, anthropology, and history, and he brings it all to bear on the landscapes of Quebec.

This part of northern Quebec, with its extensive peatlands, is a natural outdoor laboratory in which to study the postglacial ecology of the past six to seven thousand years. Peat is an excellent preservative, and trees buried in it keep their woody structure, including their annual growth rings, intact. Annual tree rings vary with climate—wider in mild years and narrower in harsh ones—yielding a unique sequence that is recognizable in all the trees from a given region. Serge Payette and his coworkers have been able to identify a continuous chronology of tree rings stretching back thousands of years in time from the present. Thus they can examine a long-dead tree trunk dug out of a deep layer of the peat and tell not only that the trunk started growing, for example, in the year 1172 and died in 1242, but also what the climate was like during that time. (I picked these dates because there is on my desk, as I write, a three-inch diameter, cut and polished section of a black spruce tree trunk [Sample M2-C6] with those exact dates determined from its rings, given to me by Serge Payette.)[2] But the preserved wood records more than just climate. For example, when a caribou steps on a branch or trunk, it leaves a characteristic scar where it stripped off the bark. When looking at a caribou scrape on ancient wood, a dendrochronologist can tell from the tree rings when the scar was produced. By counting the number of scars during a given time period in the branches and trunks taken from the peat in a fixed area, it is possible to estimate roughly the size of the caribou population that existed then, long before historical records began.[3]

These things came alive for us as we walked with Payette through his study sites, only twenty miles south of the tree line. The trees that occur here, mostly black spruce, are small and stunted. Sometimes they are ten or twenty feet high, but in exposed areas, which have less of a protective snow cover and receive the full, biting force of the winter winds, they form low, spreading mats that do not rise much above one's knees. In such a place, standing tall amidst the bonsai-like spruce trees, he began, in his melodious French-accented English, the most memorable lecture of our trip.

"We are standing in old-growth forest," he said. "Many of these trees are more than a thousand years old, some probably much older than that. By digging in the peat, we have followed individual trees back century after century. When the climate is in a comparatively warm phase, a tree will put up a regular trunk and grow tall. Then the climate becomes more severe; the trunk dies back nearly to the ground, and the tree, the same tree, resumes its matlike growth form, as you see here." One of our favorite photographs from the trip shows Serge Payette in the Ungava landscape, knapsack on his back, lecturing about old growth while the ancient black spruces—real survivors and true champion trees—spread their needled canopy around the tops of his boots.

The popular belief that the biggest, tallest trees are necessarily the oldest is easy to understand; occasionally it may be true, as in the case of redwoods and sequoias. More often, trees grow big because they are in an open field or lawn, without competitors for light, moisture, and nutrients, or because their roots are covered with a rich, fertile soil. The Kilmer Oak, a giant, spreading white oak that used to live on my campus in an agricultural field used for crop research, and which is said to have inspired Joyce Kilmer's celebrated poem, "Trees" (not my favorite poem), was found after its death in 1963 to be only 165 years old, 35 years less than the usual 200-year minimum age for "old growth." It must have grown very rapidly under favorable conditions during its comparatively short life span for a white oak. Rapid growth and huge size do not necessarily indicate old age. A tree ring expert once told me: "If I want to find the oldest tree in a region, I start looking among the little, gnarled survivors that I find clinging stubbornly to the side of a mountain, blasted by the wind and weather."

The oldest living trees that have been discovered, possibly the oldest living organisms on earth, are the bristlecone pines of California and the Great Basin, the western bristlecone pine (*Pinus longaeva*) and the Colorado bristlecone pine (*Pinus aristata*). A few of the western bristlecones, in eastern California's White Mountains, are known to be nearly five thousand years old. Yet some individual, forty-year-old "seedlings" of this species are less than six inches tall, and the height of mature bristlecones ranges from only twenty to fifty feet. These amazing trees grow near the timberline in rocky, poorly developed, phosphorusdeficient soils that receive little rainfall. There is hardly any competition from other plants in such impoverished sites, and almost no risk of fire. Thus they persist century after century, undisturbed by the invasions and catastrophes that affect bigger trees in more lush habitats. When the weather is unusually severe, they die back, losing bark and branches, but a thin rind of living tissue survives through the millennia.

Now, long after the broadcast about champion trees, I still think that the memorial grove at the Pentagon was a good idea. Other aspects of the Champion Tree Project remain problematic, however. General Cates remarked on the program that urban street trees live only an average of seven to ten years and are expensive to replace. Yet if we replant with champion trees, "perhaps that life can be extended to thirty, forty or fifty years," he said. I doubt it. Champions become champions in a particular place. If their cloned offspring are planted in a different part of the country, with a different soil and different climate, in urban environments utterly different from the ones in which they grew big, they may hardly grow at all.

More important, to portray champion trees as always the biggest of their kind, is to foster the wrong sort of image for twenty-first century America. The adulation of bigness is, to borrow a metaphor from Helena Norberg-Hodge, part of the teenage boy culture that has done so much damage to our society.[4] Big . . . fast . . . powerful—these are the attributes that most teenage boys value. Relatively few teenagers these days care about durability, safety, cost, and side effects. But an adult society cannot afford to ignore such concerns for long. There is a good reason why the replacement for the World Trade Center, whatever it may be, will not be two 110-story buildings like the ones destroyed; why no more *Concordes* are being made and none are in the offing; why an annual growth of the GNP in excess of four and a half percent is considered unhealthy by many economists; and why the consumption by this country of one-fourth of the world's energy supply is not sustainable.

Leopold Kohr wrote in *The Breakdown of Nations*: "[Bigness] is the one and only problem permeating all creation. Whenever something is wrong, something is too big." Kohr, of course, was exaggerating for effect. E. F. Schumacher expressed the sentiment more moderately and more positively in the phrase "small is beautiful."[5] If he were alive today, he might add that small is also enduring.

---

## INTIMACY WITH NATURE

When we arrived in Vancouver at the start of our vacation, the tabloid headline at the newspaper stand caught our attention. "World's Bravest Mom," it shrieked. We stopped to read. The story was simple; it needed no journalistic embellishment. Dusk, August 19, 1996. Mrs. Cindy Parolin is horseback riding with her four children in Tulameen, in southern British Columbia's Okanagan region. Without warning, a cougar springs out of the vegetation, hurtling at the neck of one of the horses. In the confusion, Steven Parolin, age six, falls off his horse and is seized by the cougar. Mrs. Parolin, armed only with a riding crop, jumps off her horse and challenges the cougar, which drops the bleeding child and springs at her. Ordering her other children to take their wounded brother and go for help, Mrs. Parolin confronts the cougar alone. By the time rescuers reach her an hour later, she is dying.

The cat, shot soon afterward, was a small one, little more than sixty pounds. Adult male cougars can weigh as much as two hundred pounds, we learn the next day from the BC Environment's pamphlet titled "Safety Guide to Cougars." We are on our way to Garibaldi Provincial Park, where we plan to do some hiking and have stopped in the park headquarters for information. "Most British Columbians live all their lives without a glimpse of a cougar, much less a confrontation with one," says the pamphlet, noting that five people have been killed by cougars in British Columbia in the past hundred years. (The number is now higher; cougar attacks have become increasingly common in the western United States and Canada in recent years.) "Seeing a cougar should be an exciting and rewarding experience, with both you and the cougar coming away unharmed."

However, the pamphlet notes, cougars seem to be attracted to children as prey, possibly because of "their high-pitched voices, small size, and erratic movements." When hiking, "make enough noise to prevent surprising a cougar . . . carry a sturdy walking stick to be used as a weapon if necessary," and "keep children close-at-hand and under control." If a cougar behaves aggressively, face it, speak loudly and firmly, wave your stick, act threateningly, fight back if attacked. Make eye contact with a cougar, the ranger behind the desk advised; on the other hand, never look straight at a grizzly bear—it may be provoked to charge. This year, he added as we turned to go, we've had a very cold, wet spring, and that has brought the cougars and bears down into the more frequented parts of the park. We should take reasonable precautions but enjoy our hiking. As he spoke, I thought of Mrs. Parolin, who had done all the things that the pamphlet recommends.

Arriving in the town of Whistler, I visited a few shops in search of a suitable walking stick, but most of the sticks I saw looked like high-tech ski poles and wouldn't have fought off an angry cricket—the one wooden stick I found was flimsy and cost $55. Besides, Joan was plainly amused at my quest, and that took the pleasure out of yuppieish shop-ping. God or nature would provide a walking stick, I decided, with a final, backward glance at the shop windows.

The next morning we got to the trailhead early. The sun was bright above the horizon—and there was a bark-free, sturdy tree branch, just the right length, that someone had left on a flat stone at the beginning of the path. We set out. Sam and Jon had been told not to run on ahead as they usually did. And for the first time in our hiking history, we asked them to make noise, even to argue if they felt like it.

The trail wound steadily upward. As soon as we crossed the park boundary, we passed from second growth to mature forest, whose majestic conifers were draped with lichen strands that caught the golden sun-shine in ten thousand ripples of light. The boys were totally silent, and I found myself thumping the ground vigorously with my stick to com-pensate. But the tiny noises I made were swallowed up in the immensity of the forest, which had its own sounds. There were Flute and Oboe Creeks, for example, which we crossed on small wooden bridges, rushing headlong down their valleys in a froth of wild, musical energy. And there were gray jays, looking, as the Peterson field guide says, like huge, over-grown chickadees, chattering loudly away as they hopped onto the boys' hands to receive cookie fragments. We met no cougars or bears, but once a vole scuttled across the path ahead.

Gradually, as we ascended, patches of bright, subalpine flowers appeared in openings in the forest. Then the patches coalesced into a dazzling, waving carpet of yellow, white, and purple, and the land opened out into a saddle between two peaks. We were in Singing Pass. The trail continued on ahead, switchbacking up the far peak, and we followed it, climbing above the tree line, past and across fields of late summer snow, which was excellent for snowballs, around a talus slope of rocks and boulders, to a point where we could look down on Russet Lake, below, or raise our eyes to a panorama of snowcapped peaks and the brilliant gleam of an arm of Overlord Glacier away in the east. We had come six miles. It was time for lunch.

The return trip was also uneventful. I pounded on rocks and roots with my stick, enjoying the totally different views from the trail now that we were going downhill. Again, there was not a sign of cougar or bear, no distinctive footprint with three lobes at the base of the pad, no rotting tree torn apart in a search for grubs. At the parking lot, I put the stick back on the flat rock. It had been totally effective in warding off danger, and I was happy to leave it for the next group that wanted to walk to Singing Pass.

The rest of our trip also passed without incident, although not without wonder. We saw old growth, western red cedars more than twelve feet in diameter in the rain forest on Meares Island. They had been saved from the chainsaws of MacMillan Bloedel by the joint action of the First Nations of Western Vancouver Island and a coalition of North American environmentalists. Their peaceful protests in 1993 resulted in 865 arrests, one of the biggest acts of civil disobedience in Canadian history. We saw magnificent coastal bogs with dwarfed yellow cedar and lodgepole pines, and with sundews like those in the New Jersey Pinelands. We watched a bald eagle feed a fully fledged offspring that was bigger than she was. But the largest carnivorous mammals we came across were three raccoons foraging at a dumpster in downtown Vancouver, two blocks from Stanley Park. No cougars at all. Not surprising, said Neil, one of our Canadian biologist friends. They had tracked an adult cougar that was wearing a radio collar for eight months while it was living in and around a suburban development, eating cats and dogs. During that entire time, it was not once seen either by the tracking crew or by any of the residents of the community. We didn't see any cougars; how many saw us we will never know.

A week after our return, I had a visit in my office from Mark, one of Joan's and my students who had just gotten back from Professor Roger Locandro's celebrated ecology class in Alaska. Mark and another student named Anna had gone for a few days of camping in Denali Park before the class started. He told me a hair-raising story about their trip. While walking alone on a ridge trail near their campsite, he heard rustling in the brush behind him; out onto the trail came a large Alaskan brown bear, followed shortly by her two, sizeable cubs. They were about twenty feet away. Nearly petrified, Mark began walking slowly backward down the trail. The cubs pushed on ahead to investigate him, but were held in check by a peremptory "Woof!" from their mother. Slowly Mark retreated and slowly the bears advanced. Mark realized that he had to get off the trail; the only practical alternative was a shale slope which ended in a precipitous drop-off about forty feet from the path. Stepping down onto the shale, he promptly fell and slid halfway to the edge, only stopping himself by wedging his foot against a rock. The cubs also left the trail and started coming down the slope toward him. They were halfway down when their mother's "Woof" stopped them in their tracks— another woof brought them quickly back up to her.

As Mark lay stretched out on the shale, he looked up and saw the six-hundred-pound mother grizzly looking down at him from the path. She was an amazing golden color, and he was intensely aware of the minute patterns and matting of her fur. He remembers that the thought of how incredibly beautiful she was filled his mind, and her great calmness overwhelmed him.

He waited ten minutes after the bear and her cubs moved on, then scrambled up to the trail and found his way back to camp. When he arrived, Anna saw that he was deathly white and his eyes were as big as saucers. She told him that while he was away she had been singing a little song—perhaps a prayer: "Bears, please don't eat us. Bears, please don't eat us." Since that time, Mark said to me, many things in his life seem less important than they used to. He feels deeply changed.

There are degrees of intimacy with nature, even if we are not always aware of them. For most of us who love wild nature, our contact with it, limited by our particular civilization, is at the surface, and our little forays only give us glimpses and hints of the immense, rolling power that lies beneath. But occasionally one of us suddenly, without warning, slips below the surface and there meets vast, impersonal forces and the probability of terrifying change—or death. Only a few seek this degree of intimacy deliberately; many of them do not survive. And yet every now and then there is that charmed individual who, against all common reason and experience, passes in and out of deepest intimacy with nature at will. I think of John Muir climbing a wildly swaying, one-hundred-foot Douglas fir in a High Sierra windstorm, only coming down hours later after the storm was over, and discovering in the ensuing calm that "never before did these noble woods appear so fresh, so joyous, so immortal."[1]

# FIVE

---

## RESTORING THE COMMUNITY

Good forms confer health upon the things that they gather together. . . . Farms, families, and communities are forms of art just as are poems, paintings, and symphonies. None of these things would exist if we did not make them. We can make them either well or poorly; this choice is another thing that we make.

Wendell Berry, *Life Is a Miracle*[1]

# THE UTOPIA FALLACY

Utopia: An imaginary or hypothetical place or state of things
considered to be perfect. . . . Latin = no-place (from Greek *ou*
not + *topos* place)

*The Shorter Oxford English Dictionary, Fifth Edition*

Do we believe in Utopia? Most would say no. But in our preoccupation
with the problems of our time, we often find it convenient to evade reality and assume that the future will take care of itself in a benevolent
way—if only we patch things up now. Sensible people who should know
better act as if the resolution of today's troubles will usher in a golden
age tomorrow, as if the ending of one evil could not possibly lead to another. This is the Utopia fallacy, and it was implicitly recognized by
early commentators, who wrote: "Pray for the welfare of the [ruling]
kingdom, for were it not for the fear of it, men would swallow one
another alive."[1]

A central problem of our age has been bigness: bigness of corporations, weapons systems, international trade organizations, human populations, media conglomerates, climate changes, and—enabling all other
bigness—the power of technology. Only the earth itself is large enough,
barely, to serve as a comparator for the scope of our undertakings—
hence the term *globalization*, used to describe the processes that have
increasingly ruled our lives during the past half-century. So relentless
has been the advance of bigness, so rapidly has it swallowed up much
that is small and self-contained, that it often appears to be about to engulf
the entire world—ending in one global production system and a few

remaining global currencies, languages, and cultures. Some fear that once this condition has been reached, historical change will cease and human freedom will be a thing of the dead past. But inevitable as it may appear, this particular vision of the future seems more threat than substance.

It is hard to imagine how globalization as presently manifested can survive. The difficulties with exports of hazardous toothpaste, pet food, children's toys, and tires from China, the serious social consequences of globalization, and the endless trade battles over agricultural subsidies are but part of the troubles that globalization makes for itself. Predictions of the end of globalization have been largely based on the extreme economic instability of the system and have come from sources as diverse as the conservative political economist John Gray, the liberal International Forum on Globalization, and the free-trade billionaires Sir James Goldsmith and George Soros.[2] The instabilities they describe are real and significant, but I believe that the end of the availability of cheap energy will be at least as important in curtailing bigness and globalization.

Elsewhere I have elaborated on the theme that the cheap energy to power globalization and expansion is a transient phenomenon largely limited to the twentieth century, and now disappearing in the twenty-first:

> From LNG [liquid natural gas] to coal gasification to oil shale to nuclear fission to breeder reactors to fusion to renewable energy, even to improvements in efficiency of energy use . . . our society looks from panacea to panacea to feed the ever-increasing demands of globalization. But no one solution or combination of solutions will suffice to meet this kind of consumption.[3]

The distinguished Canadian scientist Vaclav Smil has put this in a larger and more terrifying perspective:

> Perhaps the evolutionary imperative of our species is to ascend a ladder of ever-increasing energy throughputs, never to consider seriously any voluntary consumption limits and stay on this irrational course until it will be too late to salvage the irreplaceable

underpinnings of biospheric services that will be degraded and destroyed by our progressing use of energy and materials.[4]

For a little while there will be more conglomeration and centralization, and a few decades of additional damage—but the days of globalization seem numbered. Now is the time to prepare ourselves for the world that will likely follow, a world of increasing smallness, fragmentation, and decentralization, a world with re-empowered communities and exciting possibilities but also great problems—not Utopia.

It may seem premature to anticipate the demise of globalization while giant corporations are merging or swallowing each other up at a frightening rate, and new technologies that shrink the earth and enhance our ability to change it are announced almost daily. Yet remember the Soviet Empire in 1970: the tightness of its control over each and every individual and nation in its vast domain was unprecedented in human history. Who would have predicted that it would all be over twenty years later? The philosopher of history John Ralston Saul has written, "Nothing seems more permanent than a long-established government about to lose power, nothing more invincible than a grand army on the morning of its annihilation."[5]

After globalization comes to an end, are we likely to find large numbers of healthy communities well-stocked with public-spirited, self-sufficient women and men, communities in balance with other, similar communities, all contained within ecologically reasonable bioregions? Will we have enough good, wholesome food to eat, snug houses, roofs that don't leak, and the health to enjoy these blessings? Will we find, as Frodo did at the end of *The Lord of the Rings*, "a sweet fragrance on the air . . . the sound of singing . . . over the water," and see "the grey rain-curtain turned all to silver glass and . . . rolled back," revealing "white shores and beyond them a far green country under a swift sunrise"? I doubt it. Frodo was leaving the world in which he had struggled so long and to such good effect.[6] We will have no such choice.

We have to stay put, and it will not be Utopia. What we will find is that the damage already done to the earth's "biospheric services" (such as healthy ecosystems and a relatively stable atmosphere) and the damage to its people, plus the hidden costs of the ending of the present techno-economic system with both the bad and the good that it contains, will have conspired together to deprive us of the world of our dreams. Once the problem of bigness and globalization has faded or disappeared, other problems will take its place, just as the nineteenth-century scourge of child labor was followed by the twentieth-century scourge of idle teens haunting the street corners, drug-intoxicated and violent. Many of the troubles of the decentralized, low-tech world to come are not hard to predict, and this is the time, during the tumultuous transition, to think about them if we can. In an address delivered to the International Academy of Philosophy, meeting in Liechtenstein in 1993, Aleksandr Solzhenitsyn said: "We all see and sense that something very different is coming, something new and perhaps quite stern. No, tranquillity does not promise to descend upon our planet and will not be granted us so easily."[7] The new problems, I believe, will sort themselves out into two overlapping groups: techno-environmental problems, and human social and economic problems.

It is too early to assess the extent and permanence of the many kinds of damage done to the earth in our time, nor can we gauge the effects of this damage on our successors. For example, during the first centuries of European colonization in North America, there was rich, uncontaminated topsoil almost everywhere; abundant, easily accessible oil and iron; and apparently limitless old-growth forests containing massive trees beyond number. In the twenty-first century, these resources are sadly depleted. To fill the resource gap, we have turned to other, poorer and economically defenseless regions of the world, stripping them bare; the harm done to these places is greater than what we have inflicted on ourselves. How will we manage with what is left? A society of rich consumers is about to become a society of poor scroungers: ripping shingles, plumbing, and fixtures from abandoned suburban houses, searching the innumerable rubbish heaps left behind by a spendthrift culture for yesterday's trash—tomorrow's necessities.[8]

Some of the myriad concerns that the residents of Utopia will face are scarcity of clean drinking water, the droughts, storms, and floods brought on by grievously disrupted climates, toxic dumps everywhere, the widespread disappearance of useful, practical skills and knowledge, the spread of destructive exotic species, and poorly controlled infectious diseases. Ironically, all of this will happen in the absence of the powerful, globally coordinated science and technology that might have been counted on to give real assistance, or at least reassurance.

And what will become of the countless inventions and services created during the technological explosion once technology and its parent, science, have been starved by shortages of resources, money, and well-educated recruits? What will survive and what will disappear? Many of us will not miss supporting the army of technicians and technocrats working on an antiballistic missile defense shield that is almost certainly doomed to fail, or the production of expensive, high-tech weapons that don't win wars. The demise of NASA's hideously expensive interplanetary program and the space shuttle will not be mourned by all. These are technologies that eat up a nation's wealth while offering little that is commensurate in return, and their production requires a level of centralization, coordination, and control unique to history. In like spirit, we can gladly wave goodbye to recombinant bovine growth hormone, along with the totally unnecessary social, medical, and economic problems it has spawned.[9]

But there are technologies we may be more reluctant to lose. Computers, peripherals, and computer networks consume an inordinate amount of electric power to run and fresh, clean water and other resources to produce—far more than may be available if the majority of the world's people decide to go electronic. How much of this technology will we be willing to do without? The earth-observing satellites, both NASA and private (Landsat, EOS, Digital Globe, etc.), provide a host of vital information about climate change, sea level rise, human population shifts, and other variables that used to be very hard to measure. Similarly, we ought to wonder and worry about what it might mean to give up magnetic resonance imaging (MRI), a benign and useful but resource-consumptive technology that depends for its existence on a high order of economic centralization, exploitation, and reliable energy. The Global Positioning System (GPS—also dependent on satellites), which has revolutionized navigation, is a similar example. Many other technologies in this category seem equally vulnerable.

Perhaps under ideal circumstances we will be able to selectively save some of the softer, more beneficial technologies for use in our decentralized green Utopia. More probably, the entire technological system is so heavily interlinked and so thoroughly dependent on political and

economic props and subsidies, on a steady infusion of wealth and resources, that it will collapse like a castle made of sand when critical
parts of its infrastructure fail. Other cultures before ours have lost their
technologies. According to the ecologist Jared Diamond, the Moriori
of the South Pacific, a Polynesian people, seem to have lost their agricultural technology and reverted to a hunter-gatherer culture and a
simpler political organization when their ancestors settled the cold
Chatham Islands and found that their tropical crops would not grow.[10]
A perfectly satisfactory electrochemical fax communication system set
up between Paris and Lyons in 1865 was closed down in 1870 for lack
of public interest.[11] Have no illusions about the permanence of our particular technology.

   We should also have few illusions about society in the postglobal world.
Will we be able to build or maintain just communities with the resources
remaining? Without the wealth produced by exploitation elsewhere, will
we still be able and willing to afford the pursuit of tolerance, democracy,
and justice that we see as a necessary condition of our present lives?
Furthermore, a lesson we ought to have learned from nearly a century
of Marxist theory and practice in Russia, China, and Cambodia is that
political and social systems do not bring about fundamental changes in
human nature. That is not to say that some systems are not better than
others in certain respects; Marxist Cuba has demonstrated, for those
who are willing to look, the exciting possibilities of science and technology in the service of low-chemical, ecologically prudent agriculture.[12]
But after globalization has ended; after corporations have shrunk to
appropriate size; after we have finally admitted that biotechnology in
agriculture, with its genetically engineered food and animals, often
causes more harm than good; and after the international trade in cheap
goods produced in poor, environmentally damaged countries has been
curtailed by energy shortages and popular unrest—when the human
population has stabilized or become smaller, and semi-autonomous
communities have again proliferated, we may wonder why so many
people are still suffering.

   A major cause of suffering will be problems arising from excessive
loyalties to small communities and states. We can expect to see all the
ancient, familiar side effects of smallness and decentralization: bigotry,
racial and religious hatred, the subjugation of women, and intolerance

for handicapped people, homosexuals, visionaries, and anyone else who can be labeled as different.

E. F. Schumacher, the brilliant apostle of businesses and communities built to a manageable, human scale, defended "Balkanisation," the proliferation of small nation-states. Look, he wrote in *Small Is Beautiful*, at the example of Germany, which before Bismarck was an assortment of weak principalities and afterward a mighty (and dangerous) power. His implied question was, which is preferable?[13] But in honesty, those of us who applaud the message of Schumacher cannot help but notice the counterexample of today's Balkans—including Bosnia, Serbia, Kosovo, and Macedonia—or the fragmentation of Iraq. We can hope that after the initial chaos following the breakup of larger national entities, stabilization will occur. But always lurking in the background in a parochial world of smaller, often tribal states is the threat of local or tribal war. This is not necessarily the whiz-bang war of smart bombs, "stealth" aircraft, and super-weapons too costly to risk losing in combat. Rather, these are the wars that the military historians call "low-intensity conflicts," the little, local wars of Kalashnikovs, improvised explosive devices, suicide car bombs, kidnappings, and ethnic cleansings. Low-intensity conflict is as terrible as any war, and it is spreading rapidly, fed by regional animosities and the muscle-bound weakness of the great powers.[14] It, too, belongs to the future, a black shadow over the green Utopian countryside.

Every age has its menace. We ought not fall into the revolutionary's error of assuming that our only task is to slay *our* dragon, the dragon of centralization, bigness, and globalization. For the dragon is not entirely wicked, and many helpless people and unspoiled places are under its protection. Moreover, the very presence of this dragon has kept other dragons in hiding; they will emerge after our menace is gone. As Isaiah Berlin wrote about the ideals we pursue, "Some of the Great Goods cannot live together. That is a conceptual truth. We are doomed to choose, and every choice may entail an irreparable loss."[15] Although we have no alternative but to oppose what we know to be wrong in our days, we had best reserve some of our creative energies to help prepare those who come after us to cope with the less desirable consequences of the end of the old system.[16] In an earlier age, this is exactly what Madison, Hamilton, Jefferson, Adams, and the other framers of the U.S. Constitution and Bill of Rights did—with considerable success. They were not utopians. They built into these admirably concise documents (unlike the inflexible blueprints of Marx and Lenin) the freedom for subsequent generations to alter them according to the necessities of the times. Perhaps the original model for their farsighted flexibility was the Law of Moses, which was set forth in detail in the Hebrew Bible together with the specific instruction that it was to be interpreted and applied by the judges of future ages in accordance with the needs and conditions of their times. Similarly, America's founders knew that even after the period of colonial domination was over, there would always be other challenges for Americans to face. That they did not predict all of them does not in any way tarnish their achievement. We too must be effective realists; the starting point of realism is to be honest about the changes that the end of bigness will bring to us all.

It is time to admit that Utopia is *no+place*, and to get on with living in the real world. Our future-dominated blind faith and techno-euphoria are causing us a great deal of needless grief. They prevent us from recognizing our past successes and erode our confidence in things that already work. True, it is unreasonable to assume that every invention and way of thinking that is old is good; there are many old customs and practices that are best abandoned. On the other hand, it is equally unreasonable to ignore the evident reality that this better world of the future can be largely—but not entirely—assembled from what we already know.

---

## TRADITIONS

The Jewish holiday of Sukkot has set me thinking about ritual traditions, wondering why they assume their curious forms and how they can outlive their originators and persist for such a very long time. Sukkot, or the Feast of Tabernacles, more than any other religious holiday I observe, is defined by odd practices that make no obvious sense, yet they have been performed in approximately the same way every autumn for at least twenty-five hundred years. Although many of my friends feel that tradition sanctioned by religious authority is quite sufficient reason for observing the Sukkot rituals, I find it hard to eat a festive meal outdoors in October in a cramped and flimsy three-sided booth incompletely roofed with corn stalks, without asking myself, "Why am I doing this?" The answer, "Because your biblical ancestors spent forty years in the wilderness living in temporary booths or tents," is adequate up to a point, but it doesn't help to explain why, during six of the seven days of the festival, I am supposed to take four species of plants, a citron and an immature palm frond bound together with two branches of willow and three of myrtle, and point them successively east, south, west, and north, then down toward the earth and up toward the sky.

The scriptural origin of the four species used during Sukkot is simple enough. In Leviticus 23:40, instructions are given that beginning on the first day of the holiday, you shall take "the fruit of goodly trees, branches of palm trees, and boughs of thick trees, and willows of the brook, and you shall rejoice before the Lord your God seven days." No wasted words, no explanations. It was left to postbiblical rabbinic interpretation to fill in the needed details and standardize the ritual. Thus "the fruit of goodly trees" became the citron (*etrog* in Hebrew, one of the first citrus fruits in agriculture), and the "boughs of thick trees" were declared to be myrtle branches. The method of attachment of the willow and myrtle sprigs to the palm frond (*lulav*), using strips of palm, was specified by the rabbis, as was the notion of waving the species in different directions.

The word *sukkot* (singular *succah*), which gives rise to the name of the holiday, means booths or tabernacles in Hebrew; their biblical mention in association with the seven-day festival comes two verses after the description of the four species in Leviticus. Again, the instructions are terse, saying only "you shall dwell in booths seven days"; again, it was up to the rabbis to decide what a booth is. They decided it had to have at least three sides and be roofed with material cut from plants that grow from the ground; in modern practice this means cornstalks, bamboo poles, or prunings from many kinds of trees and shrubs. The roofing material must be spaced closely enough to provide more shade than sun, but not so closely that it keeps out the rain or obscures all the stars. A succah should not be built under a tree or under the overhang of a house or other permanent structure.

Sukkot was the outstanding Jewish festival in antiquity, so much so that it was simply referred to as *the* holiday. In late biblical times, it seems that the form of the tradition may have varied according to circumstances. In chapter eight of the book of Nehemiah, which describes events in Jerusalem in the fifth century B.C.E., following the return of some of the Jews from Babylonian exile, there is an account of the revival of the festival of Sukkot after a lapse of decades. Upon hearing a reading of the Law, presumably Leviticus, the residents went out to the hills to "fetch olive branches, and pine branches, and myrtle branches, and palm branches, and branches of thick trees, to make booths as it is written." We find no mention of either the willow or the citron, while the olive and, if the translation is correct, the pine, have been added. Evidently the Jews at that time identified the four species somewhat differently than their descendants did and do not appear to have waved them, but rather used them for the construction of the festival booths.

The question of which species were meant in the original reference in Leviticus persisted into rabbinic times. Approximately nine hundred years after the festival celebration described by Nehemiah, the rabbis of Palestine were telling a midrash, or homiletic story, in which King Solomon, the wisest of men, persuasively argues that the biblical account of the four species could refer to other plants than the citron, willow, myrtle, and even palm. Many trees bear "goodly fruit," Solomon points out—hence why the citron?—and the "boughs of thick trees" could well mean olives, not myrtle. As for the "willows of the brook," well, all trees require water. By attributing this argument to Solomon, who had been dead for a good deal more than a millennium, the rabbis were able to air their own suspicions about the identity of the four species in a politically neutral way. Having expressed their doubts, they effectively put them to rest by falling back on tradition and authority. "Who explained to Israel that the four species referred to are the citron, palm, myrtle, and willow? The sages; as it says [Proverbs], 'they are exceeding wise.'"

Thus a tradition flows through the centuries like a river, now running a straight course, now picking its way along a braided channel with side streams that peter out or perhaps rejoin the main flow at some point downstream. Meanwhile, the explanations of the tradition accumulate, like flotsam in the river, indicating as much about the character and beliefs of the communities that create the explanations as they do about the nature of the tradition itself. For example, in a modern interpretation, the tradition of Sukkot began during the days of Solomon after the

construction in Jerusalem of the First Temple. The festival, newly post-
poned until the ingathering and processing of the summer fruits and
vegetables and the wine making were finished, was lengthened to seven
days, and centralized in the city. The booths, in this explanation, were
temporary housing for the many thousands of pilgrims in Jerusalem for
the holiday. And in the twelfth century, long before the days of sociolog-
ical history and cultural anthropology, Maimonides, echoing the earlier
beliefs of Philo of Alexandria, said that the booths were to remind the
rich, in the midst of their festivities, of the tribulations of the poor.[1]

Like the booths, the four species are at the heart of Sukkot. Even
during the desperate second-century war against the Romans, the
Jewish commander Bar Kokhba made sure to supply these plants to his
soldiers so they could observe the holiday. Naturally, given their importance, many people have tried to uncover the reason why the citron,
palm, myrtle, and willow are used. Lacking an obvious biblical meaning,
the four species have accumulated more commentary than the tradition
of the booths. The fourth- and fifth-century Talmudic rabbis provided a
number of widely varying explanations. In one, the species represent, by
their shapes, parts of the human body. Thus the palm frond is like the
spine, the citron is like the heart, the myrtle leaf resembles the eye, the
willow leaf is like the mouth. Another explanation states that the four
species stand for the four matriarchs: Sarah, Rebecca, Rachel, and Leah.

Maimonides, whose matchless understanding of tradition was com-
plemented by his brilliant, rational imagination, saw the four species as
symbolic of the Jews' entry into the promised land of Israel: "To
remember this we take the fruit which is the most pleasant of the fruit of
the land, branches which smell the best, most beautiful leaves, and also
the best kinds of herbs, that is, the willows of the brook." Ever practical,
Maimonides also points out that these species were common and therefore readily available, and that they would all keep fresh for seven days.

Many of the explanations of Sukkot, especially those given by European rabbis who had no personal experience of the climate and ecology
of Israel, seem extremely far-fetched today. But early records from the
region itself make clear that the holiday and the four species were long
known to have an intimate connection with rainfall and water. This is
evident from Christian as well as Jewish sources. In John (7:37), it tells
how Jesus, on the seventh and last day of Sukkot, called out to those

who thirst—an obvious reference to the custom in that era of pouring water every day of the festival.

Arthur Schaffer, an Israeli agricultural geneticist well versed in both ecology and the historical roots of Judaism, has put together a convincing case to show that Sukkot was a water festival even more than it was a harvest celebration. Writing in a magazine appropriately called *Tradition*, Schaffer reminds us that the Israeli climate has only two seasons, a cool, wet "winter" and a hot, rainless "summer." The wet season begins in September or October, shortly after Sukkot, and continues through April. The actual amount of rain that falls varies greatly from year to year. Prayers for rain begin on the Eighth Day of Assembly, Shemini Atzeret, which follows the seventh day of Sukkot, and they are said daily until the spring festival of Passover. It is commonly assumed that these prayers are not recited during Sukkot lest they bring forth a downpour on the celebrants in their unprotected booths. Nevertheless, all of Sukkot can be seen as a preparation for the invocation of rain. In the words of Rabbi Akiba, a second-century sage martyred by the Romans, "The Holy One, blessed be He, said Pour out water before Me on Tabernacles so that your rains this year may be blessed."[2]

Above all, says Schaffer, the four species of the Sukkot ritual represent the life-giving power of water in a desert land. The Jerusalem Talmud, completed about the year 400, quotes the words of Rabbi Eliezer, "These four species grow on water therefore they come as intercessors for water." A similar comment can be found in the Babylonian Talmud, compiled about a century later. Schaffer explains that each of the species typifies a different ecology of water. The willow, a universal symbol of water, is found in streamside habitats, especially in river valleys. The myrtle grows in riverine thickets and along the slopes of stream banks on steep hillsides and mountains. The date palm is a resident of desert oases, where water bubbles up out of the ground under the rainless sky. And the citron, hardly the most delicious or useful of Middle Eastern fruits, represents irrigated landscapes, requiring more water than almost any other fruit grown at the time. Referring to the "fruit of goodly trees," the Babylonian Talmud makes a telling comparison of the Hebrew word for goodly—*hadar*—and the Greek word for water—*hydor*.

What makes traditions persist for millennia? The social historian Eric Hobsbawm has said that the modern era is replete with "invented traditions," nearly all of them related "to that comparatively recent historical innovation, the 'nation,' with its . . . nationalism . . . national symbols, histories and the rest." Thus we have flags, anthems, contests, costumes, and institutional rituals, mostly used to legitimize authority and particular value systems, or to symbolize membership in a group.[3] Can such traditions persist? Hobsbawm does not elaborate, but common sense suggests that when a nation dies or radically changes, its defining traditions should also die or radically change.

Yet the traditions of Sukkot and the four species have outlasted many nations and empires; they persist today among Jews who live in Russia, North Africa, Iran, Israel, Australia, India, and America, most of whom look and behave like the other residents of their respective countries. Why are these traditions so durable? One reason stands out in my mind. They are *organic* traditions—arising out of soil, vegetation, landscapes, and climates. Perhaps an ecologically anchored tradition can have the strength to withstand the passage of time, the rise and fall of dynasties and cultures, and even, surprisingly, transplantation to a different place on earth.

What are we to make of the explanations of our traditions? I come back to the metaphor of flotsam on a river. It can tell us much that is important to know about where the river has been and what it has encountered, but it does nothing to alter the river's flow and course. For that, other, deeper factors are at work. As the rabbis of the midrash about Solomon taught, intellect is good and necessary, but in the end a tradition has a force and mystique of its own, which one can only choose to reject or accept.

Traditions help keep communities together. The old, organic traditions, especially those that are environmentally connected and socially benign like the traditions of Sukkot, are worth maintaining for that priceless function alone. But what if traditions have been lost during centuries of turbulence? All traditions must have a beginning; why not invent new traditions if the old ones are gone? If they are created skillfully and carefully, they may develop organic attachments to the communities and the environments in which they must live.

One example is the new tradition of the Kwanzaa (or Kwanza) holiday, observed in that auspicious time for festivals, shortly after the winter solstice, between December 26 and January 1. Created in 1966 by the controversial black nationalist Maulana Karenga, while living in

California, it was first intended to give African-Americans an alternative to Christmas; although many celebrated both, and Karenga later muted his exclusive view of the holiday, referring to it as a cultural rather than religious event. The name was derived from the Swahili *kwanza*, referring to African harvest festivals of first fruits, once again demonstrating the importance in traditions of a connection to the earth.[4] One of the symbols of the holiday, reminiscent of the citron, is an ear of maize. (Maize actually originated in Mexico, not Africa; but then the citron probably was domesticated in India, not the Middle East.) Although vehemently denounced by some (mostly white) critics because of its totally invented nature and because of the stormy history of its inventor, Kwanzaa caught on quickly and is now observed peacefully and enthusiastically by millions of people.

But the origin of Kwanzaa is, ultimately, irrelevant. Indeed, many of those who observe it may not know, even so soon after its creation, where it came from. The important point about Sukkot, Kwanzaa, and similar traditions is that they transcend their creators, surviving because they sustain the members of communities in a common celebration of their mutual dependence and their deep relationship to the land.[5]

## JANE AUSTEN AND THE WORLD OF THE COMMUNITY

For two weeks, in preparation for this chapter, I wallowed in sinful luxury: rereading the six completed Jane Austen novels (especially my favorite parts), basking in the warmth and wit of her collected letters, eagerly absorbing the details of her life from her best biographies, and attentively following the arguments of her leading literary critics. I also saw the BBC and other versions of *Sense and Sensibility*, *Pride and Prejudice*, *Emma*, *Mansfield Park*, *Persuasion*, and *Northanger Abbey*, falling in love with Emma Thompson, Jennifer Ehle, Amanda Root, Sally Hawkins, and other Austen screen heroines in quick succession. Throughout—at least when I could remember to pay attention—I had two questions in mind: What does Jane Austen have to say about people, communities, and nature? And what is the cause of her resurgent popularity? Perhaps, I allowed myself to think, the questions are related.

Answering the questions proved not so simple, but I did have fun trying. I read Aunt Jane's letter, dated 8 Jan. 1817, to her nine-year-old niece Cassy, beginning:

Ym raed Yssac

I hsiw uoy a yppah wen raey. Ruoy xis snisuoc emac ereh yadret-sey, dna dah hcae a eceip fo ekac . . .[1]

I read the amusingly mordant comments she could write about her neighbors, such as the one in her letter of 3 July 1813 to her brother Francis, mentioning the "respectable, worthy, clever, agreable Mr Tho. Leigh, who has just closed a good life at the age of 79, & must have died the possesser of one of the finest Estates in England & of more worthless Nephews and Neices than any other private Man in the United Kingdoms."

I reread the last chapters of *Pride and Prejudice*, *Emma*, and *Persuasion* each three times. I read once again about Catherine Morland's cruel expulsion from Northanger Abbey, and about the ill-omened trip of Fanny Price, the Bertram sisters, and the Crawfords to the Rushworth estate, Sotherton, with its seductive, if too regularly planted, wilderness. And again I was privileged to accompany Emma Woodhouse, Miss Bates, Frank Churchill, and Mr. Knightly on the tension-charged picnic to Box Hill, surely one of the highest peaks in English literature. Yet the answers to the questions remained elusive.

As for nature, Jane Austen has very little to say directly about it. Nature in Jane Austen's works is like nature in the Hebrew Bible; it is there as a constant presence, it is an essential fact of life, and because of this it is rarely separated out for special comment. Nearly all of her surviving letters—those not destroyed after her death by her sister Cassandra—are fully taken up with recording the details of caps and gowns; the observing of births, courtships, marriages, sicknesses, and deaths; and with commenting concisely, acutely, and bitingly on the characters of the people who filled her life. True, she is keenly aware of nature. For example, in her letter of 8 Nov. 1800 to Cassandra, she writes:

We have had a dreadful storm of wind . . . which has done a great deal of mischeif among our trees. . . . I was sitting alone in the dining room, when an odd kind of crash startled me—in a moment afterwards it was repeated; I then went to the window, which I reached just in time to see the last of our two highly valued Elms descend into the Sweep!!!!! The other, . . . taking a more easterly direction sunk amongst our screen of chesnuts and firs, knocking down one spruce fir, beating off the head of another, & stripping the two corner chesnuts of several branches in its fall. . . . And what I regret more than all the rest, is that all the three Elms which grew in Hall's meadow & gave such ornament to it, are gone.

But she quickly puts this loss in unromantic perspective. "I am happy to add however that no greater Evil than the Loss of Trees has been the consequence of the storm. . . . We greive therefore in some comfort."

In *Sense and Sensibility*, the materialistic, soulless John Dashwood describes to his sister Elinor, heroine of the book, how he is improving the estate that used to be her home. On the knoll behind the house, he is putting up a greenhouse. "The old walnut trees are all come down to make room for it." Elinor keeps her feelings to herself, but we know her sense of loss. Tony Tanner, one of Jane Austen's most perceptive critics, comments on this revealing scene.

Jane Austen would not be the first person to feel that there are some trees better left standing, and some greenhouses better left unbuilt. But she was not sentimental about wildness and she recognized that society is necessarily a more-or-less continuous depredation of unchecked nature. What is implied in all her work is that human society ought to be very good indeed to justify the inroads made on "nature."[2]

As Tanner observes, Jane Austen is less directly concerned with nature than with the human society that shapes and alters it. Even so, she is well aware of the critical connection between the integrity of communities and the integrity of nature.

If they are not odes to vanished landscapes, what then is the source of the popularity of her books in these uninspired and waning days of technological oppression and commercial empire? In trying to answer this question, I soon realized that I had made a mistake in undertaking the task. How am I—a person who never watches television, who seldom goes to public meetings, who rejects polls and the idea of polling, and who mistrusts pundits who speak knowingly of the public mind—supposed to figure out why Jane Austen remains so popular? No, better stick to safe ground. I will tell you why *I* like Jane Austen.

Having made my decision, I feel not only relieved but convinced of its appropriateness. For Jane Austen would never have claimed to understand the views of the British public. That is very much a twentieth and twenty-first century sort of presumption. She knew better than anyone else the intricate rules—manners—that kept and still occasionally keep communities from unraveling, and she was as alive to the depths and subtleties of human character as anyone of her time or since; but she would not have believed that she knew what all England was thinking. On 1 April 1816 she put something like this sentiment in a letter to James Stanier Clarke, who had written to her to convey the acknowledgment and thanks of the Prince Regent for receipt of her novel *Emma*, which was dedicated to the prince. His Royal Highness had just appointed Clarke to be Chaplain and Private English Secretary to the Prince of Coburg, and, full of his own importance, he made the idiotic suggestion that Austen's next book should be a historical romance about the House of Coburg. Jane Austen's polite but firm reply follows:

> I am fully sensible that an historical romance, founded on the House of Saxe-Cobourg, might be much more to the purpose of profit or popularity than such pictures of domestic life in country villages as I deal in. But I could no more write a romance than an epic poem. I could not sit seriously down to write a serious romance under any other motive than to save my life; and if it were indispensable for me to keep it up and never relax into laughing at myself or at other people, I am sure I should be hung before I had finished the first chapter. No, I must keep to my own style and go on in my own way; and though I may never succeed again in that, I am convinced that I should totally fail in any other.

Jane Austen had two brothers who became admirals, one member of her family circle who was executed by the Committee of Public Safety in revolutionary France, and others who figured in sensational British trials, plus a working knowledge of city life. She was not a simple rustic confined by ignorance to write only about country villages. No doubt she could have expanded her artistic horizons had she chosen to do so. That she did not so choose is of course the foundation of her success and her great discovery. For her, the community was the fertile center of the world, and that was the ground in which she set her plow. She had learned and written about the types of community interaction that cause people to flourish or to founder. She knew that any tales she might write of faraway romantic princedoms, of bloody revolutions, of naval battles and urban squalor, could only be less important and less profound than the "pictures of domestic life in country villages" that were her stock in trade.

When Jane Austen delineates the dynamics of small communities, she is at her best. She makes us see and feel that the deepest foundations of civilization are anchored in the little words and deeds that pass between friends and neighbors. The best example I can find occurs during the picnic at Box Hill in *Emma*. Emma Woodhouse, confused by powerful emotions and social interactions which she is not yet mature enough to understand, makes a playfully cruel remark to the kindly but boring spinster, Miss Bates. Just a sentence, but so out of keeping with the thoughtfulness and sensitivity that have earned Emma a leading place in the community, that it shocks us like an explosion tearing across the beautiful landscape. Mr. Knightly, deeply in love with Emma, nevertheless finds it necessary to rebuke her.

> "I cannot see you acting wrong, without a remonstrance. How could you be so unfeeling to Miss Bates? How could you be so insolent in your wit to a woman of her character, age, and situation? . . . She felt your full meaning. She has talked of it since. I wish you could have heard how she talked of it—with what candour and generosity."
>
> "Oh!" cried Emma, "I know there is not a better creature in the world: but you must allow, that what is good and what is ridiculous are most unfortunately blended in her."[3]

Mr. Knightly admits the truth of this, but points out to Emma that her knowledge of Miss Bates's poverty, age, and history made this careless remark inexcusable—by implication, the community's survival depends on each of its members paying attention in their words and deeds to the known personal status of every other member, a task requiring constant, if often unconscious, effort. In her carriage on the return, "Emma felt the tears running down her cheeks almost all the way home, without being at any trouble to check them, extraordinary as they were."

I like Jane Austen because she reminds me, as Emma was reminded, of the things that ought to matter in my life, or, to put it differently, the things that would be allowed to matter in a world less dominated by superficial interactions, a world more bent on survival. It does me good to think about, even to imagine myself in, a society where there are rules of behavior that are clearly formulated and observable and that make sense. A society in which the moderate pace and human scale of life admit the possibility of my understanding what is going on around me. I like to think of a time when terrible consequences were commonly known to be firmly linked to violations of the rules that govern the relations between people in a community. In *Mansfield Park*, Sir Thomas Bertram reflects on the reason why his younger daughter is a disappointment and why his older daughter has disgraced herself and her family:

> He feared that principle, active principle, had been wanting; that they had never been properly taught to govern their inclinations and tempers by that sense of duty which can alone suffice. . . . Wretchedly did he feel, that with all the cost and care of an anxious and expensive education, he had brought up his daughters without their understanding their first duties, or his being acquainted with their character and temper.

I like Jane Austen because she describes a world in which people can make clear choices between honorable and dishonorable behavior, and can live to see that their choices matter. This is not to say that the choices in her books are all black and white. She knows, as Anne Elliot discovers in *Persuasion*, that there are "cases in which advice is good or bad only as the event decides." Nor does it mean that she thinks that dishonorable behavior always brings material punishment. Neither Wickham in *Pride and Prejudice*, nor Willoughby in *Sense and Sensibility*, nor Henry and Mary Crawford in *Mansfield Park*, nor William Elliot in *Persuasion* suffer materially from their reprehensible conduct. But they do suffer consequences, not least of which is that the community judges and rejects them when their serious flaws of character are perceived. This rejection then serves to restore the moral symmetry that the community has lost.

In the modern world, it seems harder and harder to remember what community is, or the evolved systems of speech and action needed to keep community together. Jane Austen reminds us of what we have forgotten and tells us that which we never knew. Moreover, she shows us that it is possible and necessary for us to weave durable patterns of propriety into our own lives, patterns in harmony and balance with the patterns around us.

And with all this, Jane Austen, who died at the age of forty-one of a disease that may have been treatable today, tells us that amidst the sorrows there can be constancy, love, and happy endings in the world; and that the currents of nature, the currents of community, and the currents of the soul can flow together for a while in a blissful bubbling stream that might tempt God to smile.[4]

---

## UNIVERSITIES, SCHOOLS, AND COMMUNITIES

I received an invitation to speak to the Heinz Endowments a number of years ago. This major foundation was thinking of starting a program of charitable donations to help the environment and wanted advice about how to make the best use of its money. Would I participate, the director asked, in a one-day meeting on environmental education being organized by David Orr? My topic would be the role of the university. I went, and the following is more or less what I said to Heinz.

During my first years as a board member of the Educational Foundation of America, which gives grants in a number of areas, including the environment and education, I was struck by the extreme scarcity of exciting, innovative, *useful* proposals coming out of the major research universities: Harvard, Princeton, Stanford, Berkeley, and the like. The second-tier research universities are no better; they are all scrambling to copy the bad models ahead of them.

The problems that the universities are doing little or nothing to address—either in teaching or in research—are those that we must confront if our civilization is to prosper. They are materialism in our culture; the deterioration of human communities; anomie; the commercialization (privatization) of former communal functions such as health, charity, and communication; the growth imperative; the woes of agriculture; our ignorance of the ecology of disease, especially epidemic disease; the loss of important skills and knowledge; the decline in the quality of the academic and experiential education of children; and the turning away from environmental and human realities in favor of thin, life-sucking electronic substitutes. Far from confronting these problems, universities are increasingly allying themselves with the commercial forces that are causing them. The institutions that are supposed to be generating the ideas that nourish and sustain society have abandoned this function in their quest for cash.

It is typical, for example, that with all the academics working on developing and patenting new crops, the most effective and comprehensive mechanism for monitoring and preserving the priceless and rapidly dwindling stock of *existing* crops in North America and Europe—the heritage of agriculture—was developed by a young farm couple completely outside the university system. Kent and Diane Whealy's dream was originally funded by a $6,000 foundation grant; now the Iowa-based Seed Saver's Exchange, mentioned earlier, helps gardeners and farmers all over North America and in parts of Europe exchange and grow thousands of varieties of fruits and vegetables that would otherwise have been lost because they are no longer available from commercial seed companies.[1]

Any one of the interrelated problems I've listed could inflict, or is inflicting, great damage on Western society. So why are the universities, which society counts on for answers to its most pressing problems, doing so little? It's easier to get academics at, for example, Yale and Columbia to think about processes and inventions that will bring in cash and prizes, or about abstract problems in faraway places, than it is to get them to think deeply and constructively about why large parts of New Haven and New York are so physically and socially distressed and why their students run the risk of being assaulted on the way to class. Or why, as Jonathan Kozol has told so powerfully, only a few miles from university hospitals where the rich can receive every attention, there are hospitals where desperately sick, poor Hispanics and African-Americans must spend days in corridors waiting for a room with a bed—and when they finally get one must change the bloody sheets themselves before lying down.[2] There are, of course, exceptions. Muhammad Yunus, former head of the Economics Department at Chittagong University in Bangladesh quotes, in his book *Banker to the Poor*, a blistering attack from a friend:

> You academics are failing us. You are failing in your social duties. . . . No one is accountable to anyone for anything. Certainly not you lily-white-handed academics with your cushy jobs and your jaunts abroad. You are useless all of you. Utterly useless! I am absolutely disgusted by what I see in this society. No one thinks of the poor.

Yunus turned his active concern for the poorest people in the villages around his university into a new system of making small loans within a context of social reinforcement—micro-lending—which would lift the poor, especially women, out of perpetual debt and exploitation into a position of working independence. Within a few years, Yunus's Grameen Bank had become a major force for social justice throughout Bangladesh, and its ideas had been applied in many other countries, including the United States.[3] But examples like this are vanishingly rare.

Even many ecologists, who ought to understand the importance of community interactions as a factor in survival, seem disconnected from their communities. In a paper submitted for my graduate course in conservation ecology, a student, Aviva, cited the famous statement of Rabbi Shimon the Righteous (or Shimon the Wise) in the Jewish "Ethics of the Fathers." Shimon said, "The world is sustained by three things: Torah [study of the law], labor [implying also reverence to God], and acts of loving kindness," the last of which the student translated as "community service." Her paper was an explanation and description of her unsuccessful efforts to introduce the idea of adding community service to the requirements for a graduate degree in ecology at Rutgers, a daring suggestion in today's academic world.

Aviva's education before coming to Rutgers, the state university of New Jersey, was in religious schools, from kindergarten through college. At some religious schools, including hers, a code of ethical behavior that connects the individual to the community is considered a normal and necessary part of a student's training. Although religiously mandated, this teaching is not separated from the secular curriculum; when the opportunity arises to combine the study of ethical (and other) religious laws with mathematics, biology, or literature, it is done. Of course, community service is not confined to religious schools. Many public schools, from elementary to high school, have programs that benefit their communities in a variety of ways.

When Aviva perceived a potentially useful connection between ecology and community service, she was simply using her prior education. Her experience did not prepare her for the self-centered academic climate of most secular universities, including Rutgers. True, all universities have service activities that their students can participate in, but they are generally walled off from and considered secondary to the academic curriculum. The academic curriculum itself reflects the research priorities of the specialized disciplines: if the public is part of these priorities, it is the generic, abstract public—not the folks right outside the gates. With the exception of extension service activities at land grant colleges, community service in today's universities is often the province of aging professors, a few younger, tenured mavericks, and those assistant professors who are most likely to be denied tenure when their seven years are up, because they have not brought enough research grant dollars, with the associated overhead, to the university.[4]

Universities have lost their ethical direction and sense of responsibility to the community for at least four reasons: bloated, top-down administration with its own corporate rather than academic and communal priorities; professors who are selected and promoted for their ability to bring in money, which usually goes along with a lack of commitment to the encumbering and time-wasting ideal of community service; the specialization that is thought to get grants even as it obliterates all hope of seeing and responding to the big picture; and the fact that many universities are too big and unwieldy to meet any kind of serious challenge effectively. Ironically, this race for cash is frequently unsuccessful. Whether it is a winning football team, a new gene therapy, or the promise of a genetically engineered miracle crop, the fabulous sums of money required to pursue these goals are seldom repaid in full. It is almost amusing to watch the universities use the enormous administrative and other expenses of their cash-chasing schemes as part of the justification for more capital campaigns. The Columbia University biochemist Erwin Chargaff described modern universities as "huge corporations whose only business is to lose money." He was referring to large private universities, but his remark applies to public universities as well.[5] In other words, universities mirror all too well the rest of modern culture. Under these circumstances, a foundation grant or gift to a modern university has the potential of doing more harm than good.

But you here at Heinz, I said to them, have money that the IRS says you had better give away—how should you spend it? First, identify the problems you want to concentrate on and make it clear that you are going to reward only those university grant applications that are going after the same problems. But my own prime concern would be children before they get to the university. If we don't do something on a grand scale and do it quickly, the future adults who are now four, five, six, and seven years old are going to be incapable of saving their country when they grow up, and incapable of benefiting from a rigorous college education. They need while they are young the real experience of nature, the real experience of group (community) learning and cooperation, and the real experience of the joy of sustained and satisfying work. None of these things can be done by watching television, by surfing the Net, or by absorbing the competitive values of a celebrity- and money-centered society.

This is what you might do. Remember that universities are the end stage, not the beginning, of important education. Help make kindergarten to twelfth-grade teaching an attractive profession—capable of getting and holding the cream of the crop of graduates from our best colleges, many of whom would like to teach if they thought it were a viable career. Develop ways of subsidizing the teachers of environmental studies, especially ecology and conservation, so that schools will not consider them expendable. Make sure that they have supplies, money for field trips, money for their own training, and a secure income that will not stop when the grant stops. Reward innovative, successful teachers. A society that gives bare-bones salaries to master kindergarten teachers is on shaky ground. Support university projects that get involved with children, like the Rainforest Collaborative of the University of Massachusetts/Boston, which took children from inner-city Boston schools to Massachusetts forests to learn ecological principles in the field, and then took selected children from this group, with their teachers, to Costa Rica, where they discussed the ecology of very different kinds of forests with local children. But encourage these projects to stick with the children after the initial activities are over—to interact with their parents, stay in touch with their teachers, follow the students through high school, and guide them into college.

Here are some more suggestions: Give to locally oriented projects rather than ones aimed somewhere else. Give to small institutions before you give to big ones. If you give to big institutions, don't pay any overhead unless it is earmarked for direct project support rather than an administrative rake-off, and don't pay anything unless the university chips in real dollars, too. I don't mean the phony "contributions" of space and faculty salary that would be provided anyway. All universities cry poverty, but all have secret pots of money available to give to their pet projects.

Remember that you can't do philanthropy from the office. I use the excellent example of the Educational Foundation of America, in which every solid proposal gets at least one site visit either by a trustee or by the executive director or one of her assistants. Beware of proposals that were written entirely in institutional development offices. They are easy to spot because they are all alike—the same superlatives, the same naively inflated claims, the same lack of passion and originality. You should seek out the small organizations, often just two or three people, that are doing exciting work, and help them develop proposals. If you find some of these people, you'll find others; there is a network out there.

Favor integrated proposals that involve institutions and their local communities working together, preferably including both children and adults—and make sure that they have a good chance of being self-sustaining after you end your support. It is no coincidence that many of the most successful elementary and secondary schools in the country—including those in urban ghettos—are those that have strong, organized parental support and community organizations behind them.

Don't support academic conferences, but do figure out ways of helping your grantees spread the news about successful projects so that others don't have to reinvent the wheel. Develop strategies that encourage universities to spend their money on the right things. It is essential that university administrations be cajoled, embarrassed, threatened, and, if necessary, starved into changing the reward structure so that creative, energetic professors who are more interested in teaching, community service, and important but unglamorous investigations than in big-dollar research can also be honored and promoted.

Finally, beware of us experts—too often we are people who are extremely knowledgeable about narrow systems that don't apply to the real world and never will, systems that bear no useful, time-tested relationship to the lives we are living. The great University of Wisconsin botanist and conservationist Hugh Iltis told me the following story about experts at the international corn and wheat research and breeding institute, CIMMYT, in Mexico. It was 1981; Iltis, hot and sticky from a field trip, walked into the faculty-student bar at CIMMYT for a beer. He was still carrying a bundle of annual teosinte, a Mexican grass generally believed by taxonomists to be closely related to the ancestor of corn (maize). Joining a group of advanced students and professors at a table, he threw down the teosinte, which looks nothing like corn, and asked, "What is it?" The experts gathered around, examining and muttering. Nobody knew. After a while, the waitress, a short, black-haired young woman of obvious native descent, arrived carrying a tray of beer glasses. "What is it?" Iltis asked her in Spanish, pointing to the teosinte. Glancing at the grass briefly as she set down the glasses, she answered, "*Madre de maíz*"—the mother of corn.

I concluded my talk to Heinz in the following way. Our American empire is in trouble, I said, along with the empires affiliated with it. The universities and other institutions we look to for guidance are in trouble, too; they have become dull, mean-spirited, greedy, and tired. In the past, declining empires have not recovered—they seem to gather a fatal momentum of collapse. But there are still in America and the rest of the first world vast resources of wisdom, creativity, and philanthropy, as well as the energy to use them. I see at least a half a dozen students like Aviva every year, each worth his or her weight in diamonds. And I see, outside the universities, hundreds of community-oriented organizations that could use these students' spirit, energy, and imagination—if they had the resources. Here is one way that you and other like-minded foundations can help. It will take a miracle for us to survive our crisis of spiritual and environmental health, yet miracles sometimes come to those who work and wait for them.[6]

---

## WHAT DO WE OWE OUR CHILDREN?

Shortly before the birth of our first child, Kate, now thirty-six, Joan and I bought a copy of Dr. Benjamin Spock's *Baby and Child Care*.[1] It proved to be very reassuring. A thick book with no wasted words, it told almost everything we needed to know about the coughs, the sneezes, the spots, the fevers, the crying at three in the morning, the rejection of nourishment, the emergence or nonemergence of teeth, and the terror-inspiring inclination of sleeping infants to stop breathing for a while just when their parents are getting settled in bed.

There were, of course, a few things Dr. Spock failed to mention. I knew from medical school the signs and symptoms of many rare and terrible diseases of infants. Kate got them all. Each time we would rush her to the pediatrician, Dr. Arky, and each time we had the satisfaction of bringing a little uninhibited laughter into the life of that overworked and kindly man.

Fortunately for Dr. Arky, we moved from north to central Jersey when Kate was two and found a new pediatrician, Dr. Lapkin. Although we got more experienced as time passed, and came to know Dr. Spock pretty much by heart, there was always something new. I remember, for example, my deep concern about Jonathan, our third child. When he was an infant, he did not seem to be able to hear. Joan was not particularly worried at first, but fear is contagious. What if the development of his language were affected? Anyway, it was time for his one-month visit to the pediatrician.

Dr. Lapkin appeared somewhat tired but was in a good mood; it was nearing the end of a long day. I watched as he listened to Jonathan's heart and lungs with his stethoscope and examined his reflexes.

"He looks fine," Dr. Lapkin said. "Any problems?"

"He doesn't seem to hear loud noises," I said. "I'm afraid that he may be deaf. Is there some kind of audiometer test you could give him?"

"He's responding normally to my voice as I speak," answered Dr. Lapkin; "he has been during the whole examination. Are there any other problems?" he asked, directing the question to Joan. As he left the examining room he was laughing—his laugh sounded a lot like Dr. Arky's.

Jonathan, by his teenage years, actually did show some deafness, a curious selective deafness that kept him from hearing us when we called upstairs for him to get off the phone and come wash up for dinner. His sisters and brother all had it too. It must be hereditary.

Not all visits to the pediatrician end with good news, but most do. My point is that at least for a child's early years Dr. Spock or its equivalent, plus a good pediatrician, will do fine for nearly all of the crises that a parent will face. As children grow older, however, no book, no matter how thick and complete, and no pediatrician, no matter how competent, can possibly be expected to answer every question that arises and solve every problem that comes up. Children do things that are not described in any books, unique things that you the parent must learn to cope with by yourself, without the help of paid experts.

To start with a simple example, there was the day when Kate, age five, came home from kindergarten walking oddly. She was dropped off by the carpool at the end of our very long, gravel driveway; as we watched her from the doorway coming toward us, she oscillated from side to side, singing happily. When she got closer, we were relieved to see that the cause was not medical—she was wearing only one shoe.

"Where's the other shoe?" Joan asked as she entered the house. Kate looked down at her feet, surprised.

"I don't know," she answered in a disinterested tone, "maybe in the car." Then she hobbled off to the kitchen to get a drink of orange juice.

And there was Jane, not quite four years younger than Kate. During her early years we couldn't figure out her strange lack of interest in possessing material objects. Kate's room was filled to the Plimsoll mark with bits of colored glass, pebbles, stuffed animals, seashells, wilted flowers, china figurines, and other forms of solid waste—Jane's was as empty as the Gobi desert. That empty room worried us (although Jane was not concerned). Were we neglecting her? Was she normal? On her third or fourth birthday—I've forgotten which—we made a supreme effort to overcome Jane's odd antimaterialism and presented her with a personable and charismatic stuffed elephant almost as large as she was. Jane seemed pleased, if not actually overcome with joy. There was a barely suppressed gleam in Kate's eye; she said nothing. A few days later the elephant went to live in Kate's room. We offered to escort it back, but Jane declined the offer; she had tried materialism and found it wanting.

Enjoyable as it is to record these anecdotes—some minor problems faced by parents of good children living under affluent conditions in, at that time, a country at peace—it is time to get down to business and confront the hard questions about being a parent. As children and parents age together, the questions become ever more difficult to answer. Eventually, one comprehensive question emerges, often too late for the answer to be of use: What are the responsibilities of parents in these days of social and environmental disruption and turmoil?

This is hardly a new question; every religion and culture has grappled with it. Judaism, my own tradition, is very child oriented, preoccupied with the survival of the people through its posterity and with the effects of the behavior of parents on the children. The Talmud, the principal storehouse of early rabbinic wisdom about all phases of life, a vast work of many volumes, is the obvious first place to look for a detailed compendium of the obligations of a parent. After all, the Talmudic rabbis even speculated about an arcane subject like artificial insemination in humans, which was then more than a millennium and a half from being developed. But if we search for a Talmudic equivalent of Dr. Spock, an encyclopedic list of responsibilities and operating procedures for parents, we will be disappointed. The most celebrated statement about parental responsibilities in the Babylonian Talmud is only a few sentences long. You should circumcise your son, redeem him if firstborn, teach him Torah (the law), find a wife for him, teach him a craft, and teach him to swim.[2] In rabbinic writings, when a subject of this importance is dealt with this concisely we can take it as an invitation to look beneath the surface meaning of the words.

Before beginning, however, I had better deal with the issue of sexism. The Babylonian Talmud is based on three centuries of rabbinic discussions that were begun approximately eighteen hundred years ago. Although enlightened for its time about the role and value of women, it falls far short of enlightenment by secular, twenty-first century, First World standards. For example, a father's obligations to a daughter, discussed elsewhere in the Talmud, concern only her physical need for clothing and support, plus the provision of a generous dowry so that she can marry advantageously. Fathers and mothers, sons and daughters, are treated very differently. Fortunately, there is a way around this modern difficulty that does not force us to jettison the enormously valuable Talmudic teachings. Without pretending that this was the intent of the Talmudic sages, I will simply substitute "parent" for "father" and "child" for "son" in my own consideration of the Talmudic passages. Thus I can explore the timeless wisdom of the Talmud's advice on raising children without being sidetracked by other issues.

Let's get back to the quote. Circumcision at eight days and the redemption of the firstborn son pertain only to male Jews—they are not part of an ecumenical scheme of advice to all parents, and I won't discuss them further. The same could be said of the obligation to teach the law, which in its narrower meaning is the first five books of the Hebrew Bible, but which in this case means the entire structure of Jewish law and commentary, including the Talmud. But I think that this injunction has a value even for non-Jews. It says that a parent is obliged to see that a child gets detailed, practical, ethical, instructions for relating to God (or whatever higher power one wishes to substitute), the community, and the world around.

The injunction to teach the law to one's children is underscored in the Jewish prayer known as the Sh'ma. The Sh'ma, whose name comes from the first Hebrew word of the prayer, "Listen!" is the central prayer of Jewish worship—the first prayer taught by a mother (or father) to a Jewish child and the last prayer recited by a dying person. It is said at least twice a day, morning and evening, by ritually observant Jews. The first paragraphs of the Sh'ma, taken from books 6 and 11 of Deuteronomy, contain clear instructions regarding children: "And the words which I command you this day shall be upon your heart; and you shall teach them diligently to your children, talking of them when you sit in your house, and when you walk by the way, and when you lie down, and when you rise up." In other words, education about correct religious and ethical behavior is the responsibility of the parent—it can't be delegated—and it takes time, a significant amount of the available time during waking hours. I think this obligation can be generalized to parents of any religion, and to nonreligious parents as well. In the latter case, the ethical system becomes paramount.

Some of the ethical lessons of the law are hard to teach and hard to learn. I'll give an example. Among the 613 biblical commandments that Jews recognize (of which some have not been observed since the destruction of the Second Temple in the year 70 C.E.), are several commandments that prohibit gossip about others. This applies to any idle talk about third parties. The prohibition includes favorable talk, for even friendly gossip about others can be an indirect cause of damage—for example, by provoking jealousy. But gossip is a central part of the lives of school-aged children, and they come home from school bursting with it. It is not easy to explain to a young child that it is all right to say "Billy hit me during recess," because you the parent should offer reassurance and may have to do something about it, but it is not all right to say "Billy hit Joey and Michelle during recess." As always, a parental example is the best teacher, although this is easier in the writing than in the doing. Still, the lesson can be learned. When our youngest, Sam, announced at the dinner table one night when he was ten, "One of the boys in the class argued with the teacher for five minutes about his grade, and then he threw a book at her and got sent to the office," he was aware that Joan and I really wanted to know the boy's name, but that we wouldn't ask. And we knew that he wanted to tell us who it was, but would resist the impulse. The shared feeling that comes from this kind of mutual restraint is very gratifying.

What happens if parents do or don't teach their children the law? The answer is given in the second paragraph of the Sh'ma:

If you hearken diligently to My commandments . . . I will give the rain of your land in its season, the former rain and the latter rain, that you may gather in your grain, and your wine, and your oil. And I will give grass in your fields for your cattle, and you shall eat and be satisfied. Take heed to yourselves, lest your heart be deceived, and you turn aside and serve other gods, and worship them; and the anger of the Lord be kindled against you, and He shut up the heaven, so that there shall be no rain, and the ground shall not yield her fruit; and you perish quickly from off the good land which the Lord gives you. Therefore you shall lay up these My words in your heart and in your soul. . . . And you shall teach them to your children, talking of them when you sit in your house, and when you walk by the way, and when you lie down, and when you rise up.

The question of whether an angry God withholds rain from people who violate the commandments and who don't teach them to their children is of limited interest to me. The threat may be real or it may be a metaphor; no matter. There is, however, a message in these words of the prayer that I cannot ignore, but it does not come through in English translation. One of the troubles with English is that the second-person pronoun *you* and the possessive forms *your* and *yours* are the same in the singular and the plural. This may be why the slang word *youse* and the southern regional term *you-all* or *y'all* came into being in American English; they can fill our need for a distinct second person plural. In Hebrew, however, the singular and plural forms of *you* are separate and distinct; there is no ambiguity.

In the first paragraph of the Sh'ma—the injunction to love God and teach your children God's ethical and religious laws—the *you* is singular: each parent has an obligation to teach his or her own children. In the second paragraph, which deals with rewards and punishments, there is a striking change. The *you* who receives the rain or against whom the heavens are shut is suddenly a plural. In other words, if you and you and you and the majority of other individuals who make up the community live right and teach your children to live right, then the *community* reaps the reward. If you live badly and neglect your children's spiritual and ethical education, then it is the *community* that suffers and dies.

The rabbis who first created the Sh'ma from separate passages in the Hebrew Bible were farmers or the children of farmers. They knew, as good farmers know today, that the individual cannot long survive if the community dies, but the community itself gains its strength from its individual members who live just and righteous lives and pass this way of life on to their children. As Wendell Berry wrote in *The Unsettling of America*, "One must begin in one's own life the private solutions that can only *in turn* become public solutions. . . . The use of the world is finally a personal matter, and the world can be preserved in health only by the forbearance and care of a multitude of persons."[3] These "private solutions" start with the teaching of children. As your teaching prospers, so does the world.

It is tempting to dismiss the Talmud's direction to find a wife for your son as an archaic practice that is in fact ignored by the great majority of today's Jews, including many religious Jews. But there is a modern interpretation of the teaching that bears study. We may not want or be permitted to go out and find spouses for our children, yet we can and should influence the process. We can try to define by our example and by our words what a good spouse is like; we can tell our children about the common early warning signs of trouble in potential mates. Our words may fall on deaf ears or they may arouse anger, but there is no telling when words—and examples—will be recalled at a time of need and serve their intended purpose; so the teaching cannot be neglected.

The instruction to teach your child a craft seems like the simplest item on the Talmudic rabbis' list of parental obligations. In preindustrial societies, few people survived without a craft or trade; one had to able to *do* something. The unacceptable alternatives were starvation or banditry. Today in industrial countries few parents actually teach their children a trade. There are various reasons for this: children don't expect to take over their parents' businesses, vocational or professional training has become the province of specialists, technological systems are evolving so rapidly that parents cannot keep up, and many children postpone making decisions about jobs until after they have left home. Also, the nature of work has changed radically. In the majority of today's jobs, work involves the routine application of already defined systems—there is no room for craft. Such jobs are either self-teaching or require professional training and experience. Is the advice to teach your child a trade obsolete?

My answer is no. If anything, teaching a craft to children is more important than ever; but "teaching a craft" has to be defined in contemporary terms. Except for a few, it no longer means learning the family trade from a parent. Coopers, Smiths, Wheelwrights, Bakers, Drapers, and Fletchers may carry on the family name, but not the trade that the name implies. What a parent now must teach a child about craft is altogether different from how to hold a certain tool, how to find particular materials, or how to recognize certain natural signs and signals. (It is important in children's development, however, for them to have experience using tools, handling different materials, and observing at least some of nature's useful indicators.) Rarely passed from parent to child are the little secrets that distinguish *our* barrel, *our* violin, *our* way of growing rice, from those of all others. This is sad, but there is no time for nostalgia; we have new things to teach our children about work.

First, we have to teach children that they must have a craft, even if we do not teach it ourselves. Craft, in my expanded definition, is the special skill, acquired only through training and experience, to produce particular goods and services. More than that, craft is an attitude, a way of thinking about work, which directs our energies toward the goals of continuous self-improvement, careful workmanship, pride, and completion, no matter what the task. Also implicit in the Talmudic instruction is that one's craft should make the world a better place and that it should be informed by the laws that govern our relationship with the community and the environment; otherwise, brigandage could be a craft. It is said that in the nineteenth century, Jewish teamsters in the Lithuanian city of Vilna, then a great center of Jewish learning, would suspend their physical labors at noon each day for an hour of intensive study of Torah. This is craft.

Many parents have a craft; children should see their parents at work and participate when possible. Joan is a fine example: a forest ecologist, her research is done both in the field and in the laboratory, and all of our children have watched and helped her at work at one time or another. They have had ample opportunity to observe her painstaking care, demanding standards, tireless persistence, her love of what she does, and the affection and respect her students give her. They often saw her leaving for the Pinelands or North Jersey in the morning in all weathers, dressed in her field clothes and carrying her field equipment, and they were aware that she settles down at the computer in the evening—after their bedtime when they were little. They noticed that unless she was helping them with their homework or reading one of my manuscripts, she took time off only for a quick mug of tea and a cookie at 9:00 P.M. None of our children have followed or are headed toward Joan's particular career path, but there is no question that all of them have learned craft from their mother.

Mothers and fathers, however, are not the only teachers. The endless message of television, radio, Internet ads, school, friends, and neighbors to our children is that they are valued primarily as consumers. Our message has to counter this. "Become a producer" is the theme the children should hear from us; "your worth is determined by what you are and what you do, not by what you wear or consume." Despite its overwhelming prevalence, consumerism is vulnerable. We can fight it by good example and by ridicule. Totally passive consumers are feeble, clueless, exploited creatures, easily caricatured in a way that children understand. Consumers are weak and have to be fed. Children don't want to be weak. Even if we don't teach our children a particular trade, we can teach the joy of knowing how to do useful things, how to finish what you start, how to begin to take care of your own needs, how to do things right. How to produce.

We have to teach children that a job is not necessarily a craft. Repetitive, mindless jobs producing uniform products; jobs producing shoddy, unnecessary goods; socially destructive jobs; those bureaucratic or managerial jobs that add little or nothing to real production; servile or degrading jobs; completely redundant jobs; and jobs in which one's experience, talents, and contributions are ignored are not crafts. Although a child may have to hold one of these jobs at some time in her or his life, it will still be essential to have a craft and stay in practice. We must also point out that physical labor and the work of one's hands are not inferior to intellectual work. Many of the best crafts, from surgery to teaching to farming, can combine the two.

Increasingly, schools and especially colleges are training our children for jobs that will not exist or will not be worth having after they graduate. They are being educated for the supposed benefit of "the economy," not for themselves or their communities. They are schooled in technologies and trained on equipment that will be obsolete before they get their diplomas, or they are directed toward powerful, high-tech industries that want to hire fewer, not more, employees in the years to come. An example is the communications industry: according to the Bureau of Labor Statistics, at the start of the communications revolution, between 1981 and 1988, telephone communications companies in the United States eliminated nearly 180,000 jobs, and the trend has continued unabated since then. A result of this kind of education and work experience is that it gets harder and harder to hold a job for more than a few years—a circumstance that makes craft impossible. If young children are taught the differences between craft and mere jobs, if they are taught to recognize and value the kinds of work that produce goods and services people truly need, then they have a better chance of finding satisfying, durable, productive work, even creating their own jobs, after they leave home.

Finally we get to swimming, my favorite part of the Talmud's list of instructions to parents. Because it is such a short list, the inclusion of swimming amuses me. Why swimming, of all things, in the Middle East where water is scarce and where, in ancient times, many people probably went from one year to the next without having an opportunity, let alone a need, to swim? If the Talmud had been written in Florida rather than what is now Iraq, concern about swimming would have seemed more appropriate. So this instruction is a puzzle, and in the Talmud a puzzle is an invitation to pay attention and to think.

What the Talmud is saying to a desert-dwelling people when it states that they are required to teach their children to swim is profound and important. The meaning, I believe, is this: Your children may find themselves in various kinds of environments during their lifetimes. Some of these will be quite different from the places in which they grew up. It is your paramount responsibility as a parent to teach them how to survive in different environments that they may not be able (or want) to escape or avoid. Swimming is the ultimate example of such a survival skill. (The Talmud often uses extreme examples to make a point; it is understood that less extreme cases are included in the lesson.) In water, many of the things that come naturally to us as terrestrial animals, such as walking, running, resting, breathing, eating, seeing, and hearing, will not work in the usual fashion and have to be relearned. Children relearn much more easily than adults; you must teach them to swim early.

Although the need to learn to swim is real and the Talmud's instruction can be taken at face value, swimming also serves as a metaphor. The Talmudic sages were clearly making a more general point, a point expressed as concisely as a thought in poetry. Make your children aware of the natural world around them—its many forms and manifestations, its moods and seasons, its creatures, its processes, its changes, its patterns, its predictabilities, and its unpredictability. Their healthy development, their happiness, and possibly their lives may depend on the awareness of nature that you awaken in them.

I try to teach the same lesson to my students, both undergraduate and graduate. My favorite example is my wife's former graduate student, Peter, now teaching in an American university. Peter grew up and went to college in Bulgaria, majoring in biology. He speaks Bulgarian, Russian, Spanish, and English, all fluently. He is an avid vegetable gardener. More important, in college, besides his biology education (which was thorough and advanced), he was required to learn a totally different, nonprofessional trade. Peter chose the trade of streetcar operator. Do the same, I advise my students. I don't mean you have to learn to oper- ate streetcars. But whether you are pre-med, pre-law, pre-veterinary, or pre–anything else, find a trade or craft that is totally different in which to become proficient, something that will always give you satisfaction and that will always benefit the community.

As a thoughtless economics and technology push our civilization ever faster toward some tumultuous and chaotic upheaval, teaching children to swim, in the larger sense of the instruction, takes on a special impor- tance. What is required of us is not the impossible. We do not have to be perfect, and we do not have to be prophets; we do not have to anticipate every new environment that our children may face. But as part of being attuned to nature, children should gradually be helped to understand that the world changes both spatially, which they know or guess, and with *time*, which few children comprehend. Temporal change is more rapid now than it has ever been in recorded history, and preparing chil- dren to recognize and cope with the frequently destructive changes in their environment is the swimming lesson at the start of this new cen- tury and the days ahead.

The children of wise parents will come to understand that massive and profligate use of resources, dysfunctional technologies, heedless abuse of nature, and inhumane, violence-ridden social structures pro- mote a highly unstable civilization that cannot last many more years in its present form. They will not be paralyzed by fear because they will be ready for the change. They will be accustomed to reading books rather than watching television, sorting truth from lies, playing sports rather than electronic games, conserving rather than wasting, hiking rather than riding ATVs, practicing self-sufficiency rather than helplessly consuming, building communities rather than competing alone. One of my favorite students, Aaron, an English major and a brilliant writer, helped his father with Christian Science nursing while in high school, walked the entire Appalachian Trail after high school graduation, is an experienced landscaper, has supplemented his English studies with courses in Geographic Information System and environmental planning, and is now in law school, with a special interest in environmental law and policy. I have no doubts about his ability to practice a craft that will meet the needs of the decades ahead.

Teaching children a craft is easy to talk about, hard to do. Few parents, perhaps none, will completely meet this challenge. But there is no necessity to produce Olympic swimmers, just children who are comfortable and safe in the water. Although no Dr. Spock will lead us step-by-step along this parental path, maybe we don't need anything more than a willingness to experiment, a few general guidelines like the ones in the Talmud and other sources of tested wisdom, and the ability to accept whatever consequences may flow from our best efforts. And for the lucky parents, their children will help, making the journey together a delight, whatever the ultimate destination.

---

## EPILOGUE: A CALL FOR FUSION AND REGENERATION

The German philosopher Hegel once said: "The owl of Minerva flies at dusk."[1] Minerva was the Roman goddess of wisdom, and the owl that accompanied her was her source of knowledge. What Hegel meant by saying that Minerva's owl flies at dusk is that when troubles gather and the prospects for the future of civilization seem dark and growing darker, only then, when it is too late to do anything about it, can we begin to understand what is happening. Minerva's owl is now on the wing, but there is still some daylight left, still a little time to prove Hegel wrong.

As the earth warms and the oceans rise, and as supplies of oil and gas wane while the elaborate economic and social edifice built on them begins to crumble, we find ourselves mired in a world view that offers us stark choices between two apparently opposed ways of thinking—conservative or liberal—both of which are deeply rooted in Western thought, and neither of which is capable, by itself, of coping with the unique challenges we now face. Clear and ominous portents warn us to change course, to use the wisdom we have accumulated and select from our conservative and liberal traditions the timeless, disaster-defying elements best suited for life in the different and difficult world ahead. We need a new guiding political philosophy for this age of the earth and human history; this is no time for political business as usual.

That new philosophy may now be at hand—here for the taking. Out of the centuries-long conflict between conservatism on the one hand and liberalism on the other—each side encrusted with its special truths—a new fusion of ideas is appearing, which could save humanity from a great tragedy of its own making. Our human and earth systems have reached the point when projection of present trends no longer suffices to predict the future. The point when the stretched cable suddenly snaps. The point when the great polar ice sheets slip and melt, gas tanks are empty and the power goes out, low-intensity conflicts overwhelm military forces designed for earlier kinds of war, increased stresses of living engulf long-established societies in ethnic and religious violence, and global interlinkages make all our economies vulnerable to collapse.

Amidst great environmental changes, the hundred-year span of the Age of Cheap Energy is drawing to an abrupt end, and with it the possibility for us to continue our utter dependence on the technological-economic system that we have now. Our environment, food, shelter, clothing, livelihoods, and any semblance of international stability are all at risk, and we are unprepared. People will become increasingly confused, frightened, and angry as shortages and conflicts hedge them in. The denial industry will thrive, blame will fly in all directions, scapegoating and violence will increase, law and order will be threatened. But if we look to our strengths, irrespective of their political associations, we can move toward a brighter future.

The politics and political labels of the day do not necessarily draw on the enduring ideas of the philosophies they claim to represent. Given the terrible threats we now face, we need to get beyond the contemporary political slogans of conservatism and liberalism and identify the parts of the two philosophies that are valid and useful in our rapidly changing world. As this happens, people on both sides of the political fence will find a common ground that transcends political labels without violating the tenets of either conservatism or liberalism, braiding the strands of the two philosophies into a lifeline strong enough to hold us all.

From conservatism, we can take the stability-related qualities suggested by the name conservative, which shares a root with conservation. It was the eighteenth-century conservative Edmund Burke who claimed that revolution breeds violence and societal breakdown. Durable societies evolve slowly and are held together by traditions and shared values. Revolutionary cultures lack this stabilizing influence. The subsequent excesses and anarchy of the French Revolution confirmed his predictions. From liberalism we can learn that not all traditions and values are good and useful; some may be counterproductive. Occasionally, societies must change for the welfare of their citizens. We are at such a point now.

There are, however, two problems that may impede a twenty-first century fusion of ideas taken from the conservative and liberal philosophies. These problems are inherent in the very definitions of conservatism and liberalism. Conservatism emphasizes respect for traditional practices and institutions, distrust of centralized governmental and supranational agencies, and opposition to sudden changes in the established order. Liberalism, by contrast, does not consider itself bound by traditional, established views or dogmas, nor does it necessarily respect authority and orthodox attitudes. Liberals are open to proposals for reform and new ideas for progress; they often advocate tolerance for the ideas and behaviors of minorities, regardless of how unconventional these ideas and behaviors may be.

The first problem, therefore, is how to persuade conservatives that in a changing and hazardous social and natural environment, new ideas and institutions may be needed for survival, and that strong governmental and supragovernmental agencies may be necessary to help us cope with emergency situations that transcend local or even national boundaries.

The second problem that stands in the way of a fusion of ideas from the two philosophies is how to persuade liberals to accept that some traditional, "old-fashioned" values may be essential to the maintenance of a stable society, and that these values are not in need of major alteration.

Although the basic meanings of conservatism and liberalism, as

I have defined them, are clear and relatively constant, the *political* application of the labels "conservative" and "liberal" at any given moment in history may depart considerably from the original definitions. In our time, this has been particularly true for conservatives. Conservatives in government today in the United States and other industrial nations stress a constellation of beliefs they now find acceptable: belief in the need for perpetual growth in consumption (and its attendant waste), in the freeing of individuals and corporations from governmentally imposed economic restraints, and in the conviction that the environment is not seriously or, at least, irreversibly threatened. These beliefs do not sit well with many, older, stability-related conservative values, but they are called conservative, nonetheless. Liberals, angered by the newer elements of the conservative political agenda, may be blinded to the usefulness of the other parts of conservatism, parts we have to retain. It is easy, for example, for liberal environmentalists to forget that the most environmentally farsighted president the United States has ever had was a lifelong, die-hard conservative, and that his environmental sensibilities were rooted in his conservatism. In 1907, while convening the first national conservation conference, Theodore Roosevelt wrote to the nation's governors, members of Congress, justices of the Supreme Court, industrial leaders, and newspaper editors:

> Facts, which I cannot gainsay, force me to believe that the conservation of our natural resources is the most weighty question now before the people of the United States. . . . It seems to me time for the country to take account of its natural resources, and to enquire how long they are likely to last. . . . We are prosperous now . . . we should not forget that it will be just as important to our descendants to be prosperous in their time as it is to us to be prosperous in our time.[2]

Earlier that same year, Roosevelt said, in his Arbor Day message to the schoolchildren of the United States:

> We of an older generation can get along with what we have, though with growing hardship; but in your full manhood and womanhood you will want what nature once so bountifully supplied and man so

thoughtlessly destroyed; and because of that want you will reproach us, not for what we have used, but for what we have wasted. . . . [America] must expect the penalty of the prodigal whose labor could with difficulty find him the bare means of life.[3]

These words, which apply with even more force now than one hundred years ago when they were written, can serve as a call for conservatives and liberals, in the face of national and global emergency, to find the life-preserving values upon which they can agree.

Fortunately, neither conservatism nor liberalism is as pure a philosophy as the politicians would have us believe. Each contains aspects of the other, in a mix that varies from person to person. Discovery of this commonality is the first step toward cooperation; but for this to happen, we have to identify the critical values we need to sustain a working and fair civilization in our changing environment. I will describe these values as conservative or liberal, because this is the way we usually identify them now. But none is an exclusive possession of one political group; and they may lose their older affiliations altogether as a new cooperation emerges.

Some may believe that in today's political climate and with our present system of government, conservatism and liberalism cannot be reconciled. I disagree. The conservative and liberal values that I describe below are, I believe, so compelling and obvious in their messages, so inherently appealing to a populace weary of bluster, lies, and failure, that they will be adopted once it is made clear that they can live together regardless of their political origin. This value-centered approach is very different from a prescription of specific changes and fixes (which also may be useful), but I think it has a better chance of rescuing us and our environment while there is still time.

### Conservative Values

*Patriotism* is a conservative value whose abuses and xenophobic excesses have earned it much-deserved criticism from the political left. But patriotism, in spite of Samuel Johnson's correct observation that it is the last refuge of scoundrels, also has its legitimate uses.

Patriotism can be practiced at many scales—from the local community to the nation, and perhaps even to the Earth itself. Patriotism is the admission that people who share a land, a place, and a history have a special obligation to that place and to each other.

Unfortunately, patriotism, despite the aftereffects of 9/11, seems in decline these days. Much of what now passes for patriotism is only a superficial display without commitment. Waving a flag no more makes one a patriot than watching a televised football game makes one an athlete. Patriotism does not have to involve exhibitionistic or aggressive posturing. True patriotism is active, not passive; it is incorporated into daily life, not saved for special occasions. The pioneering wildlife ecologist and conservationist Aldo Leopold wrote in 1928: "Good citizenship is the only effective patriotism, and patriotism requires less and less of making the eagle scream, but more and more of making him think."[4] Our obligation as patriots in this time of environmental and social crisis is to take back patriotism from the demagogues of the right and of the left and to promote ever-widening circles of cooperation rather than angry nationalism.

At its heart, patriotism often involves some self-sacrifice for the benefit of the larger group to which one has declared allegiance. Above all, patriotism is the opposite of selfish individualism. Military service can be and often is patriotic, but much patriotism has nothing to do with fighting. In 1910, Charles Van Hise, eminent geologist, friend to Gifford Pinchot and Theodore Roosevelt, and president of the University of Wisconsin, made this point clear in the first general conservation textbook published in the United States:

> In the days of [18]61 to '65 a million men laid aside their personal desires . . . for the good of the nation. Now it is demanded that every citizen shall . . . think not only of himself and his family, but of his neighbors, and especially of the unnumbered generations that are to follow. It is by the criterion of what is best for posterity that we should judge of the interlocking questions of economics and conservation which confront us.[5]

Because of the sacrifice that patriotism requires and the personal energy that it consumes, no patriot wants to make such a great commitment in vain. For this reason, the wise patriot remains informed about his or her community's or nation's behavior in the world around, and tries to correct this behavior if it strays from an honorable or sustainable path. In the difficult times that lie ahead—indeed that have already begun—patriotism can marshal the enormous power of thousands and millions of people working together for the common good.

When my mother, who had little taste for vegetable gardening, planted a Victory Garden during World War II and saved aluminum foil, animal fat, and rubber bands for collection, it was patriotic; she was transcending her individual needs to help her country win a just war. It is now truly patriotic to avoid buying a gas guzzler for your next car, to use compact fluorescent instead of incandescent light bulbs, to use a rake instead of a gasoline- or electric-powered leaf blower, to avoid destructive consumption, and, when possible, to buy goods or produce from local or regional producers. It is also patriotic to explain to our children why we behave in this way, and to give them opportunities to do so themselves. This contemporary patriotism has great power: it allows us, simultaneously and without internal conflict, to be patriotic to our community, our nation, and to the entire planet. This is patriotism without drawbacks.

*Thrift* is a conservative value with no long-term downside, unless it is used as an excuse for miserliness. True, in the short term, thrift can be disruptive to an economy based on a steady increase in consumer spending, which is why thrift is disparaged by many of today's self-styled conservative (and some liberal) politicians. Nevertheless, our grow-and-spend economy is going to be a short-lived affair—massive consumer and national debt, the decline of the affluent middle class, the accelerating loss of jobs in manufacturing and technology, and the ending of the flow of cheap energy that has made this economy possible are but four of the reasons why it has no future. Grow-and-spend will be replaced: to the extent that we can assist the transition to a steady-state economy, we can minimize economic collapse and suffering. Thrift is an essential component of all sustainable economies, and will be a necessary feature of postpetroleum society. The habit of thrift is best learned now, while many of us—the very ones whose thrift will have the greatest impact—can still afford it.

My parents and grandparents experienced the hardships of the Great Depression. They were afraid of debt—personal and national.

They believed in savings—not just of money, but of anything that could be used again. They distrusted frivolous spending, risky investments, and purchases on credit. They were generous, but "Do you need it?" is a question we children were frequently asked. For them, thrift was natural, if not always easy. They did not grow up in a consumer environment. It will be harder for us to be thrifty, accustomed as we are to think of ourselves as purchasers rather than producers, discarders rather than savers. Yet thrift can be a source of pride and comfort, and the benefits are very tangible. Relearning thrift may prove one of the more achievable tasks of the new political coalition.

*An appreciation of history* is a third conservative value. Of course, liberals can appreciate and invoke history, too, but their dissatisfaction with the past and their emphasis on the future often push history out of sight. An appreciation of history, environmental as well as social history, will be of enormous benefit in the days ahead. Not only can remembering the lessons of history help prevent us from repeating past mistakes, but doing so also has another function that is especially important during this period of technological, social, and environmental change. Because these changes are slow compared with the pace of daily life, happening year by year or decade by decade rather than day by day, we tend to ignore them. Thus today we tolerate new environmental and social conditions that would have horrified our grandparents and even our parents. An active awareness of history, particularly local history of the past several generations, is one of the best antidotes to the poison of unnoticed change. A sense of history can keep us from being anesthetized to the urgent need to alter our behavior to protect both traditional knowledge and our environment.

Forgetting our past inventions is a surprising problem to find in a society that worships information and technology as much as ours does. Yet once skills and knowledge are more than a few decades old, we expunge them from our historical memory no matter how useful they may have been and still might be. How to use a lathe creatively, how to sharpen a pair of scissors, how to write clearly and elegantly, how to garden without chemicals, how to feel at home in a natural environment—these are all useful, sometimes essential, parts of our historical legacy that many have forgotten. Remembering or rediscovering them is a sound, conservative strategy.[6]

*Love of place* is a conservative value that we increasingly spurn. Staying put can be hard to accept in a society where upward economic and social mobility is so often associated with exotic travel and horizontal mobility on the face of the earth. We no longer expect to live our lives in the place where we were born. Indeed, there are often valid reasons for leaving home: better jobs elsewhere, unpleasant associations, or simply the joy of seeing new faces, cultures, and landscapes. But a sense of place, even one acquired in mid- or later life, can counteract the emptiness of living everywhere and nowhere, which is so common now that travel to distant places is quick and easy and mass Western culture is all-pervasive. A sense of place allows, but does not force, you to become personally and directly familiar with neighbors and with the local environment—the plants, animals, soils, climate, and landforms where you are. This in turn can help you to live frugally, peacefully, and restoratively—valuable traits in a time of scarcity and widespread social and environmental damage.

*Community* is the final conservative value for the hard days ahead, and the one value that includes all the others. By communities, I mean more than small towns in rural settings. Neighborhoods in larger towns and cities are often communities. And beyond neighborhoods, communities can be as large as a whole city—witness the attachment its present and former residents have for the stricken city of New Orleans, or the special pride New Yorkers have in the "big apple." These and most other cities have their own, unique community cultures. Even the amorphous automobile suburbs can contain focal communities, and these are likely to grow stronger as use of the private automobile declines. Communities are groups of people who share a place and are vested in it, and who can maintain frequent, personal contact with each other. As I discussed earlier, I don't believe that electronic pseudo-communities count as communities, because there is no circumscribed place that their members share, and because the contacts are not personal in the fullest human sense. Many communities comprise people with an array of ethnic origins and religious beliefs—others are more homogeneous.

Functioning communities will be absolutely essential for a manageable transition to a stable, low-energy, low-consumption, and low-waste society. Communities can bind people together in working social groups that can cohere in the face of adversity. They can facilitate productive contact among the generations and can provide well-directed aid to the poor and handicapped. Communities offer us the life-sustaining feedback that can keep society and nature in healthy balance. This feedback loop has been heavily damaged by globalization and centralization, because critical decisions are too often made in total isolation from their far-flung consequences. Within a community, however, it is difficult to compartmentalize cause and effect of actions taken—people have a much better chance to see and experience the results of their own actions and their neighbors' actions. Communities, more than individuals, distant governments, or impersonal economic forces, can spotlight antisocial and antienvironmental behaviors, and can initiate collective action to bring this destructive behavior to a halt.

Communities can represent society at its best. They have been damaged by corporations and by the co-opting culture of television and the seductive pseudocommunities of the Internet. Yet communities persist. There are even signs that some are starting to rebound, in spite of—or perhaps in response to—the excesses of selfish individualism and the community-busting forces that characterize our era.

Communities also can represent society at its worst. Beware of the Utopia fallacy. We all know the historic drawbacks of life in tight-knit communities: parochialism, religious intolerance, racism and sexism, ignorance, limitation of physical and intellectual resources, long-lived grudges and animosities, and lack of privacy. These problems are real and are not going to go away; they will have to be faced and dealt with by the builders of postpetroleum societies, because communities are likely to regain a good deal of their autonomy in a world governed by resource scarcity, as the complex forces of centralization and amalgamation begin to weaken.[7]

Nobody can predict how soon this process will happen and how far it will extend in today's hyperglobalized, hypercomplex civilization, but I think it would be foolish not to be ready for the return of communities, warts and all. And there is one other challenge.

As communities revive, they still will exist in an interlinked world, in which the actions of one part can exert a powerful effect on the others. As the Princeton University philosopher Kwame Appiah has written: "A world in which communities are neatly hived off from one another seems no longer a serious option, if it ever was."[8] Cooperating regionally and globally to overcome widespread problems without causing conflict with local autonomy and identity will require a delicate balancing of opposing forces. Something like it was attempted, with considerable success, by the drafters of the U.S. Constitution, who combined the separate states into a federal union. We cannot know how this puzzle will be worked out during the days ahead, especially in the face of energy constraints, environmental change, and growing ethnic and religious strife. True, earthquakes, tsunamis, and other disasters have shown us that when the need is great, cooperation can overcome differences without damaging the cooperating parties. Whether this will occur when the threat is as severe but less sharply defined than an earthquake remains to be seen. Thus one of the greatest demands on our social inventiveness during the coming transition may well be to find ways to sustain our communities while at the same time strengthening those attitudes and institutions that let us coordinate life-sustaining activities on a much wider scale. Conservatives and liberals will have to use their traditions in tandem to bring this about.

## Liberal Values

*Willingness to change laws and practices* is a useful value to save from the liberal tradition and from the romanticism that is often allied with it. The enlightened idea that it is all right, in fact necessary, for a civilization to change its laws and practices to make them compatible with a drastically changed social and physical environment will be important for the future. No society can endure if it cannot adapt to altered circumstances and habitats. The current realities of our lives are global environmental change, a rapidly growing shortage of cheap energy, serious overpopulation, the many consequences of a poorly controlled technology, and a vast and growing gap between the few rich and the uncounted poor. Each of these problems interacts with the others, spawning human misery and environmental destruction on an unprecedented scale. Clearly, our economic and social systems require major adjustment. Up to now, it has been the job mostly of liberals to point this out and suggest changes.

*Environmental justice* is another important liberal value to retain. In the traditional usage, the administration of justice is the orderly and nonviolent way we relate the needs and obligations of individuals to the needs and obligations of the society in which they live. Without justice, people cannot work together effectively—and work together we must if we are to overcome the troubles we face. Thoughtful conservatives and liberals support this meaning of justice, which is mostly associated with the rights of individuals. But in the latter half of the twentieth century, a new phrase, *environmental justice*, came into existence, and this has been seen as a liberal cause. Environmental justice extends the application of justice to groups of people, especially the disadvantaged, in the context of the place in which they live. When a new, polluting waste incinerator is sited in a poor urban neighborhood, or when the removal of entire mountaintops for coal mining ruins the streams and valleys below while damaging the health and lives of local, rural residents, there are valid appeals for environmental justice to rectify the situation. The more we see people as inseparable from their environments and the more we understand that the administration of justice must encompass this unity, the better equipped we will be to cope with the growing problems of the twenty-first century—problems that are blind to the distinction between conservative and liberal.

*Reverence for nature as a role model and guide for human activities* is another liberal value we will find useful. In the late-Enlightenment liberal romanticism of Rousseau, the natural state was seen as a model for the way things ought to be, especially in human society. Rousseau's attitude toward nature was more romantic than rational; perhaps in our time it finds its counterpart in our uncritical admiration of native peoples and "natural food." Conservatives, who can be as fond of nature as anyone, have been critical of liberals who, they consider, take their nature-reverence to a foolish extreme. The pejorative term *tree hugger* is a sample of this disdain. It is true that reverence for nature can be exaggerated. But in the second half of the twentieth century and the early years of the twenty-first, the view of unmodified nature as a model of how to structure a healthy human existence has increasingly acquired a practical dimension. In case after case, the apolitical, consensus findings of modern science—especially the disciplines of ecology, toxicology, and public health—have demonstrated that it is often prudent and economical to use nature both as a role model and as a measure of health in our lives.

For example, some of the worst suffering and economic damage that occurred during the floods of 1993 and 2008 might have been avoided if we had listened to those realistic scientists who urged us to respect the natural state of the Mississippi River and stop our misguided efforts to confine it with levees and build on its floodplain. Similarly, the restoration of natural wetlands is an essential part of strategies to protect the Gulf Coast from future hurricanes like Katrina, which struck in 2005. Protected wetlands and replicas of natural wetland systems are proving economic and effective for water purification.[9]

In the case of food, it has been increasingly found that industrial agriculture and food processing have added toxic substances to the things we eat (and to the soil, air, and water that grow them)—toxins that are not there in original or "natural" systems. Thanks to the pioneering work of the ecological geneticist Wes Jackson, the ethnobotanist Gary Nabhan, and others, it has been shown that natural ecosystems—prairie, forest, desert—can serve as models for agriculture (as some prior cultures have known), providing enduring examples of more sustainable and even more productive ways of growing food than the energy-expensive, inefficient, and destructive technologies of modern chemical farming.[10] The phenomenal biological and economic success of integrated pest management, a compromise between the two approaches, is another good indicator of the power we can infuse into our own lives by an informed and appropriate respect for nature.

With the explosive development and deployment of antibiotics in the second half of the twentieth century, doctors came to believe that the age of bacterial infection was nearly over. This is what I was taught in medical school. But we celebrated too soon. It was not long before the bacteria evolved ways of disarming the antibiotics. Antibiotic-resistant infections have become the scourge of hospitals and have run rampant through our populations. Our understanding of the ecological and evolutionary give-and-take between bacteria and their hosts is now helping medical scientists design a number of new strategies of using our antibiotics and other treatments wisely. But we no longer talk about an end to bacterial infections.[11]

As we are increasingly forced to find effective ways of dealing with the consequences of floods, climate change, sea level rise, emerging global pests and diseases, and epidemics of environment-related poisonings, it should become clear that the use of nature as role model is necessary and has no political overtones.

One perennial trouble with the challenge to the status quo associated with liberalism is that taken too far it can lead to alienation from established society, encouraging romantic, sometimes irresponsible individualism and the uncritical rejection of all authority. Yet this is more a feature of the radical left than of liberalism itself. Since their origins in the eighteenth and nineteenth centuries, liberals have tended to respect authority and to work within the law to further their political aims. In this regard, liberals have more in common with traditional conservatives, while the violent rejection of authority by the radical left has its modern counterpart in the radical right (witness the destruction of the Murrah Building in Oklahoma City in 1995). Radicalism has its own philosophy and practice; it is a thing apart from the liberal or conservative political philosophies it claims to represent. Neither the violence of the left nor the violence of the right can help us now.

Perhaps this was on the mind of Courtney White, cofounder and director of the Quivira Coalition, in Santa Fe, New Mexico, when he called his remarkable new alliance of conservative ranchers and liberal environmentalists the "Radical Center." The creed of this powerful union of conservatives and liberals states: "We . . . reject the acrimony of past decades that has dominated debate over livestock grazing on public lands, for it has yielded little but hard feelings among people who are united by their common love of land and who should be natural allies."[12]

If these traditional and bitter enemies can find common ground, there is hope for a new fusion and regeneration in every region of the country and in every part of society.

The task ahead is so complex and sensitive that it will only be achieved in an atmosphere of tolerance and trust. There is going to be a transition to a new way of thinking and living. This is inevitable. It can come about through prolonged violence and misery, which is the way most such transitions occur, or it can come into existence through conscious creative effort, voluntary sacrifice, and the directed use of energy and other resources—forged by men and women, liberals and conservatives, working together. Thoughtful, responsible people will want to make the transition happen as rapidly and peacefully as possible.

None of the life-preserving conservative and liberal values listed above should be troubling to anyone who cares about civilization. We have neither the time nor resources to waste on old political debates that do not move us forward. The transition calls for a new political coalition and ideology. This is not to say that the burning issues of the nineteenth and twentieth centuries are now irrelevant. But they will have to be re-cast in a new context and examined in new ways. Some of us may find friends where we least thought to look before.

### Regenerating Communities—Beyond Conservative and Liberal

Today's communities offer plenty of examples of successful problem-solving strategies for future community-builders. A few illustrations follow. For example, in my own ethnically and religiously diverse town of approximately fourteen thousand people—Highland Park, New Jersey—there is an extensive network of volunteers who staff at least eighteen municipal committees and boards on everything from Health to Housing to Arts to Environment to Emergency Management—all coordinated by the mayor and council. The fire department and rescue squad are both staffed entirely by volunteers. The committees interact with each other, and many in the community are bound in one or more ways to this network. Difficulties arise, as is inevitable, but community networks help to resolve them. Countless communities around the country work in similar ways. Economist Hazel Henderson calls these volunteer activities part of the "Love Economy" of critical services that are unpaid, often provided by women, and which are valued globally by the United Nations Human Development Index at approximately $16 trillion per year, about half of all useful products and services.[13]

At the grassroots level, one exciting way to strengthen communities is through the revival of local currencies. An illustration is the release, in the fall of 2006, of BerkShares into the community around Great Barrington, in western Massachusetts. This project was pioneered by the late Robert Swann and by Susan Witt, executive director of the E. F. Schumacher Society. Each BerkShare costs ninety cents in U.S. dollars and is worth one dollar for purchases in participating businesses.

The 10 percent discount helps power the consumer demand for BerkShares—at the time of this writing, over a million BerkShares have been issued, with approximately 200,000 in circulation, 230 businesses signed up to accept them, and 5 banks with 11 branches are willing and eager to exchange them.[14]

The attractive bills, printed on security-quality banknote paper, bear the portraits of local historical figures. Because BerkShares cannot be spent buying merchandise from large companies like Wal-Mart, Target, and Amazon, the currency promotes shopping at locally owned stores, and the money keeps circulating within the community. Participating businesses feel that the 10 percent discount they give to customers using BerkShares is counterbalanced by increased sales and customer loyalty—it is money spent in lieu of advertising. Other municipalities, including Newark, New Jersey, with a population of more than a quarter of a million people, are studying the BerkShare system to see whether they can institute something similar. Local currency systems provide a model for a working relationship that has both conservative and liberal elements.

Communities can also promote sharing of expensive and energy-consuming equipment and services. The spectacular success of Zipcar and Flexcar is a demonstration of the opportunities available for equipment-sharing to replace ownership. These for-profit companies provide fuel-efficient rental vehicles that they park in designated lots ("home locations") throughout urban and some suburban areas in North America and Europe. Members gain access via a computerized reservation system and a wireless card that does away with keys. Hourly or daily rentals include gas, mileage, insurance, and maintenance; cars are returned to their home locations when they are no longer required. Many users have sold their personal cars, finding Zipcars and Flexcars convenient and a great deal cheaper than personal ownership of a vehicle. This is especially happening in those self-contained communities that provide jobs, shopping, and entertainment accessible by walking or by local mass transit—in other words, communities where a car is not needed every day.

A different kind of sharing can be found in several thousand communities, where the Freecycle Network, a nonprofit organization incorporated in Arizona, links up donors of surplus items—ranging from electronic equipment to furniture to musical instruments—with people in the community who can use them. This keeps serviceable items out of landfills while promoting community solidarity. There is controversy over the trademarking of the name Freecycle and over the organization's acceptance of commercial grants and advertising, but whatever form it eventually takes, the concept itself has a bright future.[15] Finally, even houses can be shared, with room rentals kept low by the exchange of services such as cleaning, cooking, and babysitting. This is being facilitated by organizations, many of them nonprofit, throughout the country.

The multiplicity of communities is also one of their strengths. There is wisdom in the old folk warning about putting all one's eggs in the same basket; it is like the biological, evolutionary wisdom of having many genetic variants within the same species. The latter has been the main driving force behind the evolution of sexual reproduction, in which the genes of two different individuals are shuffled and recombined, producing offspring both like and unlike either parent. In other words, it is important to have many different options—some may fail, some are likely to succeed. We are moving into uncharted waters. There is an advantage to having separate communities trying a variety of approaches to life in the postpetroleum age. But unlike the wastefulness and loss that is part of biological evolution, communication between individuals and organizations in different communities has the potential to redirect and rescue communities that have strayed too far along the wrong path.

I want to be very clear. Radical change in our daily lives is coming. Our most advanced thinkers are already preparing us for the changes.[16] Agriculture, education, global free trade, population growth, and suburban living are all at a watershed. Agriculture is beginning to be altered dramatically to meet the real needs of the people it feeds and the environment in which it operates. In some places, this transition will probably occur without major interruption of food supplies; it is difficult to predict which places will be so fortunate. Free trade is confronting the constraints of energy and material resource shortages and political instability; it will become smaller and less global. World population increase will come to a halt, perhaps short of predicted levels; food shortages, disease, and economic upheaval may cause a population decline. American education is failing dismally at the task of preparing students at all levels for living and working in the times ahead and is pricing itself out of existence.

Schools and universities will have to reinvent themselves—a few already are. The growth of the automobile suburbs will stop, and the existing suburbs will undergo dramatic and often destructive reorganizations. And, as I have said, local communities will regain some of the autonomy they have lost since the end of World War II. Some of these changes will be welcome, but all are likely to happen whether we want them or not. Our best hope of riding the flood without going under is to know where we are now and where we are headed in the days to come.

About thirty years ago, I was appointed chair of a committee of external examiners to evaluate a college that was in serious trouble. The fate of the college was at stake. The other committee members were also from outside universities—all were vastly senior to me. I can only assume that the reason why the president of the unlucky college chose me, a young professor, to be chair is that he thought I would be more malleable than the experienced committee members, and foolish enough to accept the job. We spent several days gathering information, then met in a conference room to draw our conclusions. The morning and part of the afternoon went by and the discussion became heated— we were at a stalemate. Time was running out; I didn't know what to do. Then the older committee member sitting next to me, a mathematician who had been dean of the faculty at one of the world's great institutes of technology, took pity and whispered in my ear. "Change the room," he said; that was all. Desperate, I called for a short break and reconvened the meeting in an adjacent room. To my complete surprise, in less than an hour we had settled all of our differences and had reached a unanimous conclusion. Now, as liberals and conservatives argue endlessly over vitally important matters that could in a different setting be resolved to the satisfaction of both sides, the dean's advice remains our best hope: It is time to change the room.

---

NOTES

The greatest part of a writer's time is spent in reading, in order to write; a man will turn over half a library to make one book.

Samuel Johnson, James Boswell's
*The Life of Samuel Johnson*

Part One: In Search of Honesty

1. John Ralston Saul, *Voltaire's Bastards: The Dictatorship of Reason in the West* (New York: Vintage Books, 1992), 175.

Pretending

1. Paul Turner, ed., Philemon Holland, trans., *Pliny's Natural History* (London: Centaur Press, 1962), 111. When Pliny correctly described ostriches sticking their heads in the shrubbery and interpreted this as an ineffectual way of hiding, he did not know that only the male ostrich does it and that he is not hiding but carefully turning the eggs he is incubating for a group of females in a communal nest on the ground within the bushes. It is just as well that Pliny didn't understand the real reason for the ostrich's behavior, or we would have lost one of the best metaphors in the language.

2. There may be more of these unfortunate exceptions in the current generation. How can children who spend most of their waking hours playing electronic games of power and destruction, in which a keystroke can wipe out an enemy, a city, or a planet, ever learn the difference between healthy and dangerous pretending? Although there is much debate over the medical consequences of electronic game addiction, one thing is clear: these game-obsessed children are not having the needed character-forming experiences in the real world, simply because they don't have the time.

3. Umberto Eco, *Travels in Hyperreality* (San Diego, CA: Harcourt Brace Jovanovich, 1986). See the title essay in this collection, which is closely tied to the theme of pretending in America.

4. Richard Lewontin, "The DNA Era: Cellular Complexity and the Failures of Genetic Engineering," *GeneWatch* 16 (July/Aug. 2003): 3–7.

5. Richard Lewontin, *It Ain't Necessarily So: The Dream of the Human Genome and Other Illusions* (New York: New York Review Books, 2000), 147, 151, 163.

6. Sheldon Krimsky, *Science in the Private Interest: Has the Lure of Profits Corrupted Biomedical Research?* (Lanham, MD: Rowman and Littlefield, 2003), 111–112.

7. Mae-Wan Ho, *Genetic Engineering: Dream or Nightmare?* 2nd ed. (Dublin: Gateway, 1999), 108, 162.

8. M. Tepfer, "Risk Assessment of Virus-Resistant Transgenic Plants," *Annual Review of Phytopathology* 40 (2002): 467–491; M. Tepfer et al., "Transient Expression in Mammalian Cells of Transgenes Transcribed from the Cauliflower Mosaic Virus 35S Promoter," *Environmental Biosafety Research* 3 (2004): 91–97.

9. E. Finkel, "Engineered Mouse Virus Spurs Bioweapon Fears," *Science* 291 (2001): 585.

10. National Research Council, *Biological Confinement of Genetically Engineered Organisms* (Washington, DC: National Academy Press, 2004).

11. T. Clarke, "Pest Resistance Feared as Farmers Flout Rules," *Nature* 424 (2003): 116; C. Macilwain, "U.S. Launches Probe into Sales of Unapproved Transgenic Corn," *Nature* 434 (2005): 423; Dan Charles, "U.S. Courts Say Transgenic Crops Need Tighter Scrutiny," *Science* 315 (2007): 1069.

12. Rudolf Jaenisch and Ian Wilmut, "Don't Clone Humans!" *Science* 291 (2001): 2552.

13. Peter Rosset, "Taking Seriously the Claim that Genetic Engineering Could End Hunger," in *Engineering the Farm: Ethical and Social Aspects of Agricultural Biotechnology*, ed. Britt Bailey and Marc Lappé (Washington, DC: Island Press, 2002), 81–82, 86.

14. There are serious animal health, human health, and social consequences of the use of recombinant bovine growth hormone. These range from an increase in bovine mastitis and contaminated milk to the spread of antibiotic-resistant bacteria to added economic pressures on small dairy farmers. They and other issues are discussed in my article "Unethical Contexts for Ethical Questions," in *Expanding Horizons in Bioethics*, ed. A. W. Galston and C. Z. Peppard (Dordrecht, Netherlands: Springer, 2005), 19–34.

15. Vandana Shiva, "Globalization and the War against Farmers and the Land," in *The Essential Agrarian Reader: The Future of Culture, Community, and the Land*, ed. Norman Wirzba (Lexington: University Press of Kentucky, 2003), 133.

16. Charles M. Benbrook, "Who Controls and Who Will Benefit from Plant Genomics?" Presented at "The 2000 Genomic Seminar: Genomic Revolution in the Fields: Facing the Needs of the New Millennium," American Association for the Advancement of Science Annual Meeting, Washington,

DC, Feb. 19, 2000. Available at http://www.biotech-info.net/AAASgen
.html (accessed July 12, 2007).

17. Miguel A. Altieri, *Genetic Engineering in Agriculture: The Myths, Environmental Risks, and Alternatives*, 2nd ed. (Oakland, CA: Food First, 2004), 9.

18. M. Pollan, "The Great Yellow Hype," *New York Times Magazine*, Mar. 4, 2001, 15–16.

19. Altieri, *Genetic Engineering*, 12–13.

20. H. Sudarshan, "Foreword," in Vanaja Ramprasad, *Hidden Harvests: Community Based Biodiversity Conservation* (Bangalore: Green Foundation; Seattle, WA: Edmonds Institute, 2002), 5.

21. Orrin H. Pilkey and Linda Pilkey-Jarvis, *Useless Arithmetic: Why Environmental Scientists Can't Predict the Future* (New York: Columbia University Press, 2007), 62, 64, 65, 137, 138.

22. William J. Broad, "Missile Test Success! (Next Time We'll Hit the Target)," *New York Times*, April 18, 1999, section 4, p. 5; Frances FitzGerald, "The Poseurs of Missile Defense," *New York Times*, June 4, 2000, section 4, p. 19; Charles Seife, "Report Says Early Strikes Can't Shoot Down Missiles," *Science* 301 (2003): 287.

23. Philip E. Coyle III, "U.S. Missile Defenses in Europe: The Putin Alternative," *Defense Monitor* 36 (July/Aug. 2007): 2–3.

24. Theodore A. Postol, "Why Missile Defense Won't Work," *Technology Review* 105 (April 2002): 44; Gary Taubes, "Postol v. the Pentagon," *Technology Review* 105 (April 2002): 52–59.

25. Ariana Huffington, "The Penalty Was Steep for a Missile Defense Whistle-Blower," *Los Angeles Times*, March 15, 2002; www.commondreams.org/views02/0315-06.htm (accessed April 28, 2008).

26. David Ehrenfeld, *The Arrogance of Humanism*, rev. ed. (New York: Oxford University Press, 1981), 125.

27. A number of cities and towns in the United States and Canada, including Los Angeles and Vancouver, have banned or strictly regulated leaf blowers, but resistance to the bans from some landscapers and property owners has been considerable.

## Brainstorming Has Its Limits

1. From a telephone conversation between the author and Wendell Berry.

2. Mary Midgley, "Why Smartness Is Not Enough," in *Rethinking the Curriculum*, ed. Mary E. Clark and Sandra Wawrytko (Westport, CT: Greenwood Press, 1990), 39–52. Those who wish to explore further Midgley's profound

understanding of the powers and limitations of modern science and the human mind will enjoy her books, including: *Science as Salvation: A Modern Myth and Its Meaning* (London: Routledge, 1992); and *Evolution as a Religion: Strange Hopes and Stranger Fears* (London: Methuen, 1985). Similar issues are explored in the following books: David Ehrenfeld, *The Arrogance of Humanism*, rev. ed. (New York: Oxford University Press, 1981); note especially the chapter titled "Emotion and Reason" and David W. Orr, *Earth in Mind: On Education, Environment, and the Human Prospect* (Washington, DC: Island Press, 1994); most of the book is relevant, particularly the chapter titled "Some Thoughts on Intelligence."

## Nothing Simple

1. Andrew C. Revkin, "Report Tallies Hidden Costs of Human Assault on Nature," *New York Times*, April 5, 2005, F2; Jonathan A. Foley et al. "Global Consequences of Land Use," *Science* 309 (2005): 570–574.

2. Michael Balter, "The Baby Deficit," *Science* 312 (2006): 1894–1897.

3. Garrett Hardin, "Nobody Ever Dies of Overpopulation," in *Stalking the Wild Taboo*, 2nd ed. (Los Altos, CA: William Kaufmann, 1978).

4. Herman E. Daly, *Beyond Growth: The Economics of Sustainable Development* (Boston: Beacon Press, 1996).

5. Jianguo Liu and Jared Diamond, "China's Environment in a Globalizing World," *Nature* 435 (2005): 1179–1186.

6. Mark Elvin, *The Retreat of the Elephants: An Environmental History of China* (New Haven: Yale University Press, 2004). A remarkable work: a literary, cultural, historical, and environmental masterpiece, which tells the long, not always simple history of the Chinese people and their relationship to their land.

7. I. Miyazaki, *China's Examination Hell: The Civil Service Examinations of Imperial China*, trans. Conrad Schirokauer (New York: Weatherhill, 1976), 14.

8. Charles O. Hucker, *China's Imperial Past: An Introduction to Chinese History and Culture* (Stanford, CA: Stanford University Press, 1975), 10.

## The Comforts of Fantasy

1. I chose the following mystery novels because I enjoyed them and because they are representative of the best work of the authors cited in this chapter:

Marjory Allingham, *Police at the Funeral* (Harmondsworth, U.K.: Penguin, 1984); Stephanie Barron, *Jane and the Unpleasantness at Scargrave Manor* (New York: Bantam Books, 1996); Agatha Christie, *The Mysterious Affair at Styles* (New York: Bantam, 1961); Nicolas Freeling, *Sand Castles* (New York: Warner Books, 1989) R. Austin Freeman, *Mr. Pottermack's Oversight* (Mineola, NY: Dover, 1985); Jonathan Gash, *The Sleepers of Erin* (Harmondsworth, U.K.: Penguin, 1984); Tony Hillerman, *A Thief of Time* (New York: Harper and Row, 1990); P. D. James, *An Unsuitable Job for a Woman* (New York: Popular Library, 1972); H. R. F. Keating, *Bats Fly Up for Inspector Ghote* (Chicago: Academy, 1974); Donna Leon, *Blood from a Stone* (New York: Penguin, 2006); James Melville, *A Sort of Samurai* (New York: St. Martin's Press, 1981); Ellis Peters, *One Corpse Too Many* (New York: Fawcett Crest, 1979); Dorothy L. Sayers, *The Nine Tailors* (New York: Harcourt, Brace and World, [1934] 1962); Rex Stout, *In the Best Families* (New York: Bantam Books, 1962); Robert Van Gulik, *Necklace and Calabash* (New York: Charles Scribner's Sons, 1967).

## Part Two: Keeping Track of Our Losses

1. Joseph A. Tainter, *The Collapse of Complex Societies* (Cambridge: Cambridge University Press, 1988), 1.

## Rejecting Gifts

1. Neil Postman, *Building a Bridge to the Eighteenth Century: How the Past Can Improve Our Future* (New York: Alfred A. Knopf, 1999), 6.
2. William Carlos Williams, *Paterson* (New York: New Directions, 1951).
3. Gene Logsdon, "The Importance of Traditional Farming Practices for a Sustainable Modern Agriculture," in *Meeting the Expectations of the Land: Essays in Sustainable Agriculture and Stewardship*, ed. Wes Jackson, Wendell Berry, and Bruce Coleman (San Francisco: North Point Press, 1984), 3–18.
4. David Ehrenfeld, "Forgetting," in *Beginning Again: People and Nature in the New Millennium* (New York: Oxford University Press, [1993] 1995), 65–72. In this chapter, I discuss the problem of the loss of knowledge and skills that is the silent partner of the technological advances of the past half-century.

## The Uses and Risks of Adaptation

1. Archie Carr, *Ulendo: Travels of a Naturalist In and Out of Africa* (New York: Knopf, 1964), xi–xii.
2. George Orwell, *The Road to Wigan Pier* (New York: Harcourt, [1937] 1972), 207.
3. Edward R. Tufte, *The Cognitive Style of PowerPoint* (Cheshire, CT: Graphics Press, 2003), 3, 4. Tufte is best known for his book *The Visual Display of Quantitative Information* (Cheshire, CT: Graphics Press, 1983), a landmark exposition on how to present graphs, charts, maps, and tables effectively, without losing important information.
4. Clive Thompson, "PowerPoint Makes You Dumb," *New York Times Magazine*, December 14, 2003, 88.
5. David Leonhardt, "Technology Eases the Ride to Higher Tolls," *New York Times*, July 4, 2007, C1, C4.

## When Machines Replace People

1. In Karel Capek's play *R. U. R.: (Rossum's Universal Robots): A Fantastic Melodrama* (trans. Paul Selver [Garden City, NY: Doubleday, Page, 1923]), Capek, who coined the word *robot*, lamented the folly of replacing humans with machines. The ending is happier than what we are likely to experience. For another take on the same theme—the preciousness of the human spirit—see P. D. James's *The Children of Men* (Toronto: Penguin Books, 1992).

## Pseudocommunities

1. Bruce Wilshire, *The Moral Collapse of the University: Professionalism, Purity, and Alienation* (Albany: State University of New York Press, 1990), 9.
2. Lewis Mumford, *The City in History: Its Origins, Its Transformations, and Its Prospects* (New York: Harcourt, Brace and World, 1961); see especially pp. 511–513 ("Families in Space") for an explanation of why modern America has proved such fertile soil for pseudocommunities.
3. Jane Austen, *Emma* [1815], vol. 4 of *The Oxford Illustrated Jane Austen*, 3rd ed., ed. R. W. Chapman and M. Lascelles (Oxford: Oxford University Press, 1988) (also see the citations for the chapter titled "Jane Austen and the World of the Community"); Elizabeth Gaskell, *North and South* (London:

Penguin Books, [1854–55], 1970); *Wives and Daughters* (London: Penguin Books, [1864–66] 1986). Elizabeth Gaskell, like Jane Austen who preceded her, had an exceptional ability to describe the society and environment around her. Both of these books record changes in traditional community life; the first because of the new power of industrial development, and the second because of the growing importance of science.

4. Peter W. Huber and Mark Mills, "Dig More Coal—The PCs Are Coming," *Forbes* (31 May 1999), 70–72. Many people think of the Internet and e-mail as free, but as the authors point out, it takes "about 1 pound of coal to create, package, store and move 2 megabytes of data. The digital age, it turns out, is very energy-intensive" (70).

5. Bill McKibben, *The Age of Missing Information* (New York: Plume, 1993); George W. S. Trow, *Within the Context of No Context* (New York: Atlantic Monthly Press, [1981] 1997); Clifford Stoll, *High-Tech Heretic* (New York: Anchor Books/Random House, 1999). (The three books cited here are important reading for anyone who thinks that electronic pseudocommunities are the way of the future.)

6. Norman H. Nie and D. S. Hillygus, "The Impact of Internet Use on Sociability: Time-Diary Findings," *IT and Society* 1 (2002): 1–20. The principal finding of this extensive study was: "On average, the more time spent on the Internet at home the less time spent with friends, family and on social activities; in contrast, Internet use at work has little effect on sociability" (11). These authors conclude that the Internet, when used at home, displaces social contact—hardly a surprising finding. Another example of the many papers on this subject is Eric J. Moody, "Internet Use and Its Relationship to Loneliness," *CyberPsychology and Behavior* 4 (2001): 393–401. But for a more cautious view, see Robert D. Putnam, *Bowling Alone: The Collapse and Revival of American Community* (New York: Simon and Schuster, 2000). For Putnam, who also cites extensive research, television is more of a villain than the Internet. He states, "It is precisely those Americans most marked by . . . dependence on televised entertainment who were most likely to have dropped out of civic and social life—who spent less time with friends, were less involved in community organizations, and were less likely to participate in public affairs." He qualifies his assessment somewhat by noting that "we cannot be entirely certain that [heavy users of television and other electronic entertainment] would be more sociable in the absence of television" (246).

7. Trip Gabriel, "Computers Help Unite Campuses But Also Drive Some Students Apart," *New York Times*, Nov. 11, 1996.

8. H. F. Harlow and R. R. Zimmerman, "Affectational Responses in the Infant Monkey," *Science* 130 (1959): 421–432.

## Obsolescence

1. A complete list of references on the causes of the dinosaurs' extinction would probably be longer than this book. Here are a few, listed in chronological order, that I have found interesting: Editor, "How Geomagnetic Reversals May Wipe Out Animals," *New Scientist* (Jan. 29, 1976): 231; Dale A. Russell, "The Gradual Decline of the Dinosaurs—Fact Or Fallacy?" *Nature* 307 (1984): 360–361; J. L. Cloudsley-Thompson, "The Success of the Dinosaurs," *New Scientist* (May 24, 1984): 13–17; Virginia Morell, "How Lethal Was the K-T Impact?" *Science* 261 (1993): 1518–1519; David Archibald, "Were Dinosaurs Born Losers?" *New Scientist* (Feb. 13, 1993): 24–32; Editor, "What Really Killed the Dinosaurs?" *New Scientist* (Aug. 16, 1997): 23–27; Charles R. Marshall, "Mass Extinction Probed," *Nature* 392 (1998): 17–20; Richard A. Kerr, "No 'Darkness at Noon' to Do In the Dinosaurs?" *Science* 295 (2002): 1445, 1447; Kate Ravilious, "Killer Blow," *New Scientist* (May 4, 2002): 28–31; Arturo Casadevall, "Fungal Virulence, Vertebrate Endothermy, and Dinosaur Extinction: Is There a Connection?" *Fungal Genetics and Biology* 42 (2005): 98–106.

2. George Sturt, *The Wheelwright's Shop* (Cambridge: Cambridge University Press, [1923] 1993), 202. George Sturt, who also wrote under the name of George Bourne, wrote some of the best early twentieth-century descriptions of the impact of technological change—and the "meaningless" work that came with it—on the lives of people and their communities; see George Sturt [George Bourne, pseud.], *Change in the Village* (New York: Augustus M. Kelley, [1912] 1969). Two other works that provide great insights into change and obsolescence are: Christopher Lasch, *The True and Only Heaven: Progress and Its Critics* (New York: W. W. Norton, 1991); and Kirkpatrick Sale, *Rebels against the Future: The Luddites and Their War on the Industrial Revolution: Lessons for the Computer Age* (Reading, MA: Addison-Wesley, 1995).

3. Jeff Rothenberg, "Ensuring the Longevity of Digital Documents," *Scientific American* (Jan. 1995), 44–45.

4. Wendell Berry, *The Unsettling of America: Culture and Agriculture* (San Francisco: Sierra Club Books, 1977), 59.

5. Stephen Jay Gould, *Wonderful Life: The Burgess Shale and the Nature of History* (New York: Norton, 1989); one of the best, most thoroughly readable, and fascinating books on evolutionary history. Gould understood the danger of making sweeping generalizations involving such concepts as obsolescence.

## Accelerating Social Evolution

1. John Ralston Saul, *Voltaire's Bastards: The Dictatorship of Reason in the West* (New York: Vintage, 1992), 64.

2. Scott Adams, *The Dilbert Principle: A Cubicle's-Eye View of Bosses, Meetings, Management Fads and Other Workplace Afflictions* (New York: HarperCollins, 1996), 198. Dilbert seems to have struck a nerve among modern corporate Americans: a quick Google search for "Dilbert," made in July 2007 found 12,800,000 matches. Perhaps the best and funniest commentary on the curious and frequently self-destructive structure and (mal)function of modern organizations was written long before Dilbert by the naval historian C. Northcote Parkinson: C. Northcote Parkinson, *Parkinson's Law: And Other Studies in Administration* (Boston: Houghton Mifflin, 1957).

3. Paul De Palma, "http://www.when_is_enough_enough?.com," in *The Best American Science and Nature Writing 2000*, ed. David Quammen (Boston: Houghton Mifflin, 2000), 66–67.

4. Erwin Chargaff, *Heraclitean Fire: Sketches from a Life before Nature* (New York: Rockefeller University Press, 1978), 167.

## Writing

1. P. G. Wodehouse, *The Most of P. G. Wodehouse* (New York: Simon and Schuster, 1960); *Right Ho, Jeeves* (Harmondsworth, UK: Penguin, 1953). There is no point in criticizing bad writing without providing examples of how writing ought to be done. Wodehouse produced nearly one hundred books during his long life, all elegantly written and eminently readable. *Right Ho, Jeeves* is one of his best and funniest; the other volume listed includes some of his finest short stories.

2. For other examples, see Richard Mitchell, *The Graves of Academe* (Boston: Little, Brown, 1981). Mitchell is one of our most insightful, amusing—and ruthless—critics of bad writing.

3. As instant messaging has been supplemented by the highly abbreviated text messaging and its variants, my students' writing seems to have deteriorated further. But at least the affect-effect distinction no longer confuses them. Now, people have abandoned *affect* entirely, in favor of *impact* used as a verb, which cannot be confused with *effect*. This has impacted me greatly, partly because I am reminded of wisdom teeth when I hear it, and partly because the usefully nuanced neutrality of *affect* has given way to the

heavy-handed bludgeoning of *impact*. See Sven Birkerts, *The Gutenberg Elegies: The Fate of Reading in an Electronic Age* (New York: Fawcett Columbine, 1994). Many people have claimed that modern problems with writing are a direct result of the demise of reading among a population hooked on TV, the Internet, and electronic communication. Birkerts is among the most eloquent and persuasive of them. See also: Nicholas Carr, "Is Google Making Us Stupid?" *Atlantic* 302 (July/August 2008), 56–63.

4. George Orwell, "Politics and the English Language," in *The Collected Essays, Journalism and Letters of George Orwell, IV: In Front of Your Nose, 1945–1950*, ed. Sonia Orwell and Ian Angus (New York: Harcourt Brace Jovanovich, 1968), 127–140. This is the classic essay on the connection between writing and life.

5. Improving one's writing is a lifelong affair, but it need not be onerous or dull. Two readily available books on writing that I have found useful and fun to read are Strunk and White, a slim classic which has gone through several editions, and Garner, a more contemporary and much more comprehensive work, well on its way to becoming a classic: William Strunk Jr. and E. B. White, *The Elements of Style*, 3rd ed. (New York: Macmillan, 1979); Bryan A. Garner, *Garner's Modern American Usage* (New York: Oxford University Press, 2003). Because Garner distinguishes between and describes both American and British usage, his book is useful beyond North America.

## Part Three: Toward a Sustainable Economics

1. Wendell Berry, *Life Is a Miracle: An Essay against Modern Superstition* (Washington, DC: Counterpoint, 2000), 54.

## Affluence and Austerity

1. Paul L. Wachtel, *The Poverty of Affluence: A Psychological Portrait of the American Way of Life* (Philadelphia: New Society, 1989).

2. Michael N. Dobkowski and Isidor Wallimann, eds., *The Coming Age of Scarcity: Preventing Mass Death and Genocide in the Twenty-first Century* (Syracuse, NY: Syracuse University Press, 1998); Sir James Goldsmith, *The Trap* (New York: Carroll and Graf, 1994); James Howard Kunstler, *The Long Emergency: Surviving the End of Oil, Climate Change, and Other Converging Catastrophes of the Twenty-First Century* (New York: Grove Press, 2006); Matthew R. Simmons,

*Twilight in the Desert: The Coming Saudi Oil Shock and the World Economy* (New York: John Wiley and Sons, 2005).

3. Harry Wiland and Dale Bell, *Edens Lost and Found: How Ordinary Citizens Are Restoring Our Great American Cities* (White River Junction, VT: Chelsea Green, 2006).

## Energy and Friendly Fire

1. David Pimentel and Tad W. Patzek, "Ethanol Production Using Corn, Switchgrass, and Wood; Biodiesel Production Using Soybean and Sunflower," *Natural Resources Research* 14 (March 2005): 65–76. Pimentel was one of the first scientists to document, in 1973, the enormous energy costs of modern agriculture (corn), and Patzek is a world authority on energy, peak oil, and alternative fuels. Here, the authors show that the production of ethanol from corn, switchgrass, and wood, and the production of biodiesel from soybeans and sunflowers require more energy inputs than the fuels, themselves, provide. To claim that these technologies will solve the energy crisis is another example of pretending on a grand scale.

2. Mathis Wackernagel and William Rees, *Our Ecological Footprint: Reducing Human Impact on the Earth* (Philadelphia: New Society Publishers, 1996), 27, 127. The concept of the "ecological footprint"—the amount of land needed to provide the energy and materials and to absorb the wastes of an average person living in a particular country—has become widely used since the publication of this immensely readable and amusingly illustrated little book. The ecological footprint of the United States is calculated by the authors to be 5.1 hectares (12.6 acres) per person. The authors, Canadians, show that most industrialized countries have an "ecological deficit"—in other words, they require much more land than they actually have to meet their consumption and waste assimilation needs. The deficit is made up by the nonindustrialized nations.

3. P. H. Tyedmers, R. Watson, and D. Pauly, "Fueling Global Fishing Fleets," *Ambio* 34 (2005): 635–638.

4. Pimentel, D., R. Zuniga, and D. Morrison, "Update on the Environmental and Economic Costs Associated with Alien-Invasive Species in the United States," *Ecological Economics* 52 (2005): 273–288.

5. Angus Wright, *The Death of Ramón González: The Modern Agricultural Dilemma*, rev. ed. (Austin: University of Texas Press, 2005). This classic work, first published in 1990, describes the impact of export agribusiness, with its

heavy dependence on chemical pesticides, on the lives of the Mixtec and Zapotec peoples of Mexico. Wright skillfully uses his knowledge of history, politics, Green Revolution agricultural technology, American foreign development policy, and the personal lives of Mexican native peoples to tell the story of a tragedy that is being repeated in many countries around the world.

6. A. G. Murray, R. J. Smith, and R. M. Stagg, "Shipping and the Spread of Infectious Salmon Anemia in Scottish Aquaculture," *Emerging Infectious Diseases* 8 (Jan. 2002): 1–5; "Special Section: Effects of Mobile Fishing Gear on Marine Benthos" [a series of articles on the effects of trawling and dredging on the marine seafloor], *Conservation Biology* 12 (Dec. 1998): 1178–1240.

7. M. Lavoie, "Impact of Global Change and Forest Management on Carbon Sequestration in Northern Forested Peatlands," *Environmental Reviews* 13 (2005): 199–240.

8. David Ehrenfeld, "The Joseph Strategy: Transforming the Cheap Energy Economy," *Orion* 22 (Sept./Oct. 2003): 18–28.

9. Vaclav Havel, "Our Moral Footprint," *New York Times*, Sept. 27, 2007, A33.

10. See the final chapter, "Epilogue: A Call for Fusion and Regeneration." For the state of global oil and gas supplies, see www.peakoil.net, the Web site for the Association for the Study of Peak Oil and Gas, an international organization of scientists and others, founded by the distinguished British petroleum geologist Colin Campbell (accessed May 19, 2008).

## Durable Goods

1. Herbert C. Kraft, *The Lenape: Archaeology, History, and Ethnography* (Newark: New Jersey Historical Society, 1986).

2. K. David Harrison, *When Languages Die: The Extinction of the World's Languages and the Erosion of Human Knowledge* (New York: Oxford University Press, 2007).

3. John R. Quinn, *Fields of Sun and Grass: An Artist's Journal of the New Jersey Meadowlands* (New Brunswick, NJ: Rutgers University Press, 1997), 71–72. Quinn has woven art, natural history, and human history together into a beautifully illustrated description of the meadowlands and its inhabitants, over time.

## Preserving Our Capital

1. David Ehrenfeld, "The Joseph Strategy: Transforming the Cheap Energy Economy," *Orion* 22 (Sept./Oct. 2003): 18–28. In this article, I explain that it is still possible to relieve the energy crisis and save the best parts of industrial civilization. I begin with the story of Joseph, who used a combination of capital (grain) accumulation and strict rationing to maintain Egypt through years of famine. In today's energy crisis, which also translates into food shortage, two additional strategies—technological innovation and reduction of energy consumption—are available. Stockpiling of oil for more than a few months' to a year's supply is impractical; rationing and technological innovation are likely to be important but have significant associated problems; only reduction of energy consumption is fail-safe and guaranteed to work.

2. S. M. Alexanyan and V. I. Krivchenko, "Vavilov Institute Scientists Heroically Preserve World Plant Genetic Resources Collections during World War II Siege of Leningrad," in *Seed Savers 1992 Summer Edition*, Decorah, IA. See "Seed Savers Exchange" in citations for the chapter "I Reinvent Agriculture."

3. At the time of this writing, in 2008, the U.S. federal excise tax on gasoline was 18.4 cents per gallon, unchanged since 1993, a small percentage of the total cost per gallon, which ranged from about $2.70 to $4.00. (At the same time, state gas taxes ranged from approximately 8 cents per gallon in Georgia and Alaska to 32 cents in New York and Wisconsin.) In Iran, the total price per gallon was 38 cents, which was causing the country to experience severe economic problems and gasoline shortages. And in England, gas was more than $7.00 per gallon, of which half was tax. Some American economists and even a few politicians were suggesting that the federal gas tax be raised to 50 cents or one dollar, which would probably help to decrease our gas consumption and our dependence on foreign oil (by causing consumers to buy more gas-efficient cars and reduce their driving). But the chances of this wartime-style increase being approved by Congress and the president, without more tangible evidence of crisis, seemed nil.

4. Albert A. Bartlett, "Forgotten Fundamentals of the Energy Crisis," *American Journal of Physics* 46 (1978): 876–888. One of the most important scientific papers of the past half century, it is understandable by any well-educated person, including those who, like me, do not have special training in physics. For an update by Bartlett of this famous article see "Reflections on Sustainability, Population Growth, and the Environment—Revisited," *Renewable Resources Journal* 15 (Winter 1997–98): 6–23.

5. Mike Davis, "Home-Front Ecology: What Our Grandparents Can Teach Us about Saving the World," *Sierra* 92 (July/Aug. 2007): 52, 54, 71.


### Conservation for Profit

1. James A. Spotila, *Sea Turtles: A Complete Guide to Their Biology, Behavior, and Conservation* (Baltimore, MD: Johns Hopkins University Press, 2004); an authoritative, readable, and beautifully illustrated book.
2. Anne B. Meylan and David Ehrenfeld, "Conservation of Marine Turtles," in *Turtle Conservation*, ed. Michael W. Klemens (Washington, DC: Smithsonian Institution Press, 2000), 96–125. This article includes an extensive discussion of the history and side effects of turtle farming operations, with many references. It is the primary source of information on sea turtle farming supplied in this chapter.
3. Shi Haitao, James F. Parham, Michael Lau, and Chen Tien-Hsi, "Farming Endangered Turtles to Extinction in China," *Conservation Biology* 21 (Feb. 2007): 5.
4. Editor, "A Question of Breeding," *Nature* 449 (2007): 2.
5. Jerry Guo, "Year of the Tiger," *Nature* 449 (2007): 16–18. For a more complete discussion of the tiger-farming controversy, see Virginia Morell, "Can the Wild Tiger Survive?" *Science* 317 (2007): 1312–1314.
6. Valerius Geist, "How Markets in Wildlife Meat and Parts, and the Sale of Hunting Privileges, Jeopardize Wildlife Conservation," *Conservation Biology* 2 (Mar. 1988): 15–26; Candace K. Mathiason et al., "Infectious Prions in the Saliva and Blood of Deer with Chronic Wasting Disease," *Science* 314 (2006): 133–136.
7. Gordon Grigg, "Kangaroo Harvesting and the Conservation of Arid and Semi-Arid Rangelands," *Conservation Biology* 3 (June 1989): 194–197.
8. H. Goodwin et al., "Tourism, Conservation and Sustainable Development: Case Studies from Asia and Africa," International Institute for Environment and Development, *Wildlife and Development Series No. 12*, 1998.
9. R. Macpherson, "Coral Reefs Still in Danger from Tourism," *Science* 317 (2007): 1498–1500.
10. K. A. Alexander et al., "*Mycobacterium tuberculosis*: An Emerging Disease of Free-Ranging Wildlife," *Emerging Infectious Diseases* 8 (June 2002): 598–601.
11. Rodolfo Vasquez and Alwyn H. Gentry, "Use and Misuse of Forest-Harvested Fruits in the Iquitos Area," *Conservation Biology* 3 (Dec. 1989): 350–361.

12. Charles M. Peters et al., "Oligarchic Forests of Economic Plants in Amazonia: Utilization and Conservation of an Important Tropical Resource," *Conservation Biology* 3 (Dec. 1989): 341–349.

13. Michael J. Balick and Robert Mendelsohn, "Assessing the Economic Value of Traditional Medicines from Tropical Rain Forests," *Conservation Biology* 6 (Mar. 1992): 128–130.

14. Douglas J. McCauley, "Selling Out on Nature," *Nature* 443 (2006): 27–28. Counterarguments by several correspondents and McCauley's response can be found in *Nature* 443 (2006): 749–750.

## Hot Spots and the Globalization of Conservation

1. A. P. Dobson, J. P. Rodriguez, W. M. Roberts, and D. S. Wilcove, "Geographic Distribution of Endangered Species in the United States," *Science* 275 (1997): 550–553.

2. J. Dewey Soper, "Solitudes of the Arctic," *Canadian Geographic Journal* 7 (1933), 105, 113; "The Lake Harbour Region, Baffin Island," *Geographical Review* 26 (1936): 426–438.

3. Ibid.

4. Citations are for this quotation and those in the paragraphs following. Norman Myers et al., "Biodiversity Hotspots for Conservation Priorities," *Nature* 403 (2000): 853–858; Stuart L. Pimm and Peter Raven, "Extinction by Numbers," *Nature* 403 (2000): 843–845; William K. Stevens, "The 'Hot Spot' Approach to Saving Species," *New York Times*, March 14, 2000, F3.

5. Thomas L. Friedman, "Foreign Affairs; Saving the Lost World," *New York Times*, March 31, 2000, A27.

6. It should be pointed out that Conservation International works in many ways besides entering into agreements of the sort described here. In the Venezuelan territory of the Yek'wana and other indigenous peoples, for example, they are opposing destructive development projects and the expansion of agriculture into the forest reserve and are working effectively to improve the economic conditions of the native residents in a sustainable, environmentally sound manner.

7. Herman E. Daly and John B. Cobb Jr., *For the Common Good: Redirecting the Economy Toward Community, the Environment, and a Sustainable Future* (Boston: Beacon Press, 1989), 234.

8. J. P. Rodriguez et al., "Globalization of Conservation: A View from the South," *Science* 317 (2007): 755–756.

9. Barbara Kingsolver, *The Poisonwood Bible* (New York: HarperCollins, [1998], 1999), 520.

10. Wes Jackson, *Becoming Native to This Place* (Lexington: University Press of Kentucky, 1994), 99. Jackson is a superb example of those modern scientists who are making their science benefit the world and the place where they live.

11. Wendell Berry, *The Unsettling of America: Culture and Agriculture* (San Francisco: Sierra Club Books, 1977), 26; a classic account of what has gone wrong in the three-way relationship among modern society, the individual, and the earth.

12. Jackson, *Becoming Native to This Place*, 97; Wendell Berry, "Conservation Is Good Work," *Amicus Journal* 14 (1): 1992.

13. Christopher Uhl, "Conservation Biology in Your Own Front Yard," *Conservation Biology* 12 (Dec. 1998): 1177.

14. Frank K. Lake, "In the Sacred High Country," in *Intricate Homeland: Collected Writings from the Klamath-Siskiyou*, ed. Susan Cross (Ashland, OR: Headwaters Press, 2000), 86.

## Putting a Value on Nature

1. Robert Costanza et al., "The Value of the World's Ecosystem Services and Natural Capital," *Nature* 387 (1997): 253–260.

2. William K. Stevens, "How Much Is Nature Worth? For You, $33 Trillion," *New York Times*, May 20, 1997, C1, C5.

## The Downside of Corporate Immortality

1. George P. Marsh, *Man and Nature: or Physical Geography As Modified by Human Action* (New York: Charles Scribner, 1864); *The Earth as Modified by Human Action: A New Edition of Man and Nature* (New York: Arno Press, [1874] 1970). Epigraph from the 1864 edition, 55. This is the first modern, ecologically oriented book on conservation, written by a remarkable linguist, historian, lawyer, architect, diplomat, and geographer who predicted many of the most important environmental issues of the twentieth and twenty-first centuries.

2. Richard Grossman, "Can Corporations Be Accountable?" (in two parts), *Rachel's Environment and Health Weekly* (July 30, Aug. 6, 1998): nos. 609, 610; "Corporate Rights vs. Human Need," *Rachel's Environment and Health Weekly* (Nov. 18, 1999): no. 677. *Rachel's* newsletter, now known as *Rachel's Democracy and Health News*—a major source of reliable, timely environmental and

health news—is published by the Environmental Research Foundation, P.O. Box 160, New Brunswick, NJ, 08903, dhn@rachel.org.

3. Chief Justice John Marshall, "Trustees of Dartmouth College v. Woodward, 4 Wheaton, 518, 1819," in *Documents of American History*, 5th ed., ed. Henry Steele Commager (New York: Appleton-Century-Crofts, 1949), 221–222.

4. George P. Marsh, *The Earth as Modified by Human Action: A New Edition of Man and Nature* (New York: Scribner, Armstrong, 1874), 53–54.

5. Michael Paulson, "WTO Case File: the Venezuelan Oil Case," *Seattle Post-Intelligencer*, Nov. 22, 1999; Editor, "WTO and Sea Turtles Clash Again and Again," *Animal Welfare Institute Quarterly* 50 (Summer 2001); Editor, "EU Complies with WTO Ruling on Hormone Beef, Calls on U.S. and Canada to Lift Trade Sanctions," *European Union News Release No. 61/03* (Oct. 2003).

6. Miki Tanikawa, "Mitsubishi Admits to Broad Cover-Up of Auto Defects," *New York Times*, Aug. 23, 2000, A1, C19.

7. Keith Bradsher, "Documents Portray Tire Debacle as a Story of Lost Opportunities," *New York Times*, Sept. 11, 2000, A1, A23. Also see "Questions Raised about Ford Explorer's Margin of Safety," *New York Times*, Sept. 16, 2000, C1, C4.

8. Joseph Borkin, *The Crime and Punishment of I. G. Farben* (New York: Free Press, 1978). Unbelievable if it weren't true; written by a senior Nuremberg war trials prosecutor.

9. Stanley Falkow and Donald Kennedy, "Antibiotics, Animals, and People—Again!" *Science* 291 (2001): 397. This editorial in *Science* states that "using the same antibiotics in people and animals is a bad idea"; Ellen K. Silbergeld and Polly Walker, "What If Cipro Stopped Working?" *New York Times*, Nov. 3, 2001, A23; Marc Kaufman, "Ending Battle with FDA, Bayer Withdraws Poultry Antibiotic," *Washington Post*, Sept. 9, 2005, A03.

10. For a discussion of the legal aspects of corporate death see: Mary K. Ramirez, "The Science Fiction of Corporate Criminal Liability: Containing the Machine Through the Corporate Death Penalty," 47 *Arizona Law Review* 933 (Winter, 2005)

Part Four: Relating to Nature in a Human-Dominated World

1. Thomas Berry, *The Dream of the Earth* (San Francisco: Sierra Club Books, 1988), 215; a brilliant book by a historian of cultures who is both an advocate of sustainable communities and America's dean of religiously inspired environmentalists.

## Wilderness as Teacher

1. E. F. Schumacher, "Buddhist Economics," in *Small Is Beautiful: Economics as if People Mattered* (New York: Harper and Row, 1973), 50–58.

## Death of a Plastic Palm

1. The following two books, mentioned in the chapter, will not help you with diseases of plastic trees; however, if you have real ones to care for they will be of use. Richard Harris, *Arboriculture: Care of Trees, Shrubs, and Vines in the Landscape* (Englewood Cliffs, NJ: Prentice-Hall, 1983); John R. Hartman, Thomas P. Pirone, and Mary Ann Sall, *Pirone's Tree Maintenance*, 7th ed. (New York: Oxford University Press, 2000).

2. George Orwell, *The Road to Wigan Pier* (New York: Harcourt Brace Jovanovich, [1937] 1972); *Coming Up for Air* (New York: Harcourt Brace Jovanovich, [1939] 1969); "Some Thoughts on the Common Toad," in *The Collected Essays, Journalism and Letters of George Orwell*, vol. 4 (New York: Harcourt Brace Jovanovich, [1946] 1968). Orwell was not only the most perceptive social critic of his time, but also a competent naturalist. His comments on the substitution of plastic and steel for nature have not been surpassed.

## Scientific Discoveries and Nature's Mysteries

1. Peter Berthold et al., "Rapid Microevolution of Migratory Behaviour in a Wild Bird Species," *Nature* 360 (1992): 668–670. A commentary on this paper by William J. Sutherland, "Genes Map the Migratory Route," appears in the same issue of *Nature*, 625–626.

2. Peter Berthold, *Bird Migration: A General Survey* (New York: Oxford University Press, 1993); an authoritative and readable account, with a good summary of Perdeck's experiment with starlings, 140–142, and Papi's work with odor cues, 163–165.

3. Ibid.

4. Gabrielle Nevitt, "Foraging by Seabirds on an Olfactory Landscape," *American Scientist* 87 (Jan.–Feb. 1999): 46–53.

5. C. Walcott, J. L. Gould, and J. L. Kirschvink, "Pigeons Have Magnets," *Science* 205 (1979): 1027–1029; Thomas H. Maugh II, "Magnetic Navigation an Attractive Possibility," *Science* 215 (1982): 1492–1493; Kathryn Brown,

"Animal Magnetism Guides Migration," *Science* 294 (2001): 283–284; Kenneth J. Lohmann et al., "Regional Magnetic Fields as Navigational Markers for Sea Turtles," *Science* 294 (2001): 364–368; Thomas Alerstam, "The Lobster Navigators," *Nature* 421 (2003): 27–28; Larry C. Boles and Kenneth J. Lohmann, "True Navigation and Magnetic Maps in Spiny Lobsters," *Nature* 421 (2003): 60–63.

6. Michael Reilly, "To Get Home, a Pigeon Just Follows Its Nose," *New Scientist* 191 (Aug. 5, 2006): 13.

7. Charles Darwin, "Origins of Certain Instincts," *Nature* 7 (1873): 418.

8. Henry Walter Bates, *The Naturalist on the River Amazons: A Record of Adventures, Habits of Animals, Sketches of Brazilian and Indian Life, and Aspects of Nature under the Equator, during Eleven Years of Travel* (New York: Viking Penguin [1863] 1989), 222–223.

9. Editor, "Bar-tailed Godwit Is King of the Skies," *New Scientist* 185 (Mar. 26, 2005): 21.

10. C. Groot, "On the Orientation of Young Sockeye Salmon (*Oncorhynchus nerka*) during Their Seaward Migration Out of Lakes," *Behaviour*, Supplement 14 (Leiden: E. J. Brill, 1965).

## I Reinvent Agriculture

1. For up-to-date, interesting, and practical information about vegetable and fruit gardening with heirloom, nonhybrid seeds, as well as access to a staggering number of these vegetable varieties, you can't do better than join the Seed Savers Exchange (3094 North Winn Road, Decorah, Iowa 52101; 563-382-5990; www.seedsavers.org), which was cited in the chapter "Spending our Capital." There is a modest annual fee, for which you will be richly rewarded. Seed Savers Exchange is the major nongovernmental, nonprofit organization in North America dedicated to preserving our heritage of vegetable varieties. SSE is also active in northern Europe. There are also numerous nonprofit and for-profit seed companies that carry heirloom vegetable varieties. Many of these companies are excellent, and I enjoy reading their catalogs on a dark, February evening. A few of the organizations that I like are: Johnny's Selected Seeds, Maine; Seeds of Change, New Mexico; Territorial Seed Co., Oregon; Southern Exposure Seed Exchange, Virginia; Fedco Seeds, Maine; Native Seeds/SEARCH, Arizona; and Nichols Garden Nursery, Oregon. There are scores of other U.S. and Canadian seed companies, equally good—nearly all are listed in

the *Garden Seed Inventory* (6th ed., 2004) available from Seed Savers Exchange. Try to pick at least one seed grower from your own geographic region.

2. John Wyndham, *The Day of the Triffids* (Harmondsworth, U.K.: Penguin, 1954).

## Thinking about Breeds and Species

1. Niko Tinbergen, *The Herring Gull's World* (New York: Basic Books, 1960).

2. Wendell Berry, *Life Is a Miracle: An Essay against Modern Superstition* (Washington, DC: Counterpoint, 2000).

3. The fellow at Yale may have carried his argument to an absurd extreme, but it was not entirely without basis. More recently, in her doctoral dissertation (at Brown University, 2007) on cultural attitudes to invasive species, my friend Wei Ying Wong has pointed out that we have to "move away from the simplistic good-bad dualism that currently pervades the discourse of invasion ecology." This makes good sense and is especially important when we are speaking about invasives to people with different values, languages, and cultures. Having heard Wei Ying present her ideas to ecologists, first at an invasive species biology meeting in Beijing and then at my university, in New Jersey, I was impressed to see how enthusiastically they were endorsed by most scientists in her audiences. For good discussions of the biology of invasive species, I recommend the following books: Julie L. Lockwood, Martha Hoopes, and Michael Marchetti, *Invasion Ecology* (Malden, MA: Blackwell, 2007). H. A. Mooney et al., eds., *Invasive Alien Species: A New Synthesis* (Washington, DC: Island Press, 2005). Charles S. Elton, *The Ecology of Invasions by Animals and Plants* (London: Methuen, 1958). (This is the first great book on the subject, and perhaps still the best, written by one of the finest ecologists of the twentieth century. The two concluding chapters on conservation are classics.); Alexandre Meinesz, *Killer Algae: The True Tale of a Biological Invasion* (Chicago: University of Chicago Press, 1999). (The biology and politics of a devastating species invasion. It reads like a mystery story, but it is definitely not fiction.)

4. Ernst W. Mayr, *Evolution and the Diversity of Life: Selected Essays*. (Cambridge: Harvard University Press [1976] 1997). Everything you might want to know about what a species is, from one of the greatest evolutionary biologists of our time. He remained productive until his death in 2005, at age one hundred. In places the book can be difficult for laymen.

## Strangers in Our Own Land

1. Oliver R. W. Pergams and Patricia A. Zaradic, "Is Love of Nature in the US Becoming Love of Electronic Media? 16-Year Downtrend in National Park Visits Explained by Watching Movies, Playing Video Games, Internet Use, and Oil Prices," *Journal of Environmental Management* 80 (2006): 387, 393.

2. Richard Louv, *Last Child in the Woods: Saving Our Children from Nature-Deficit Disorder* (Chapel Hill, NC: Algonquin Books of Chapel Hill, 2005). In his carefully researched, informative, and compelling book, Louv describes the essentially modern problem of "nature-deficit disorder" among our children, and suggests ways of correcting it. Gary Paul Nabhan and Stephen A. Trimble, *The Geography of Childhood: Why Children Need Wild Places* (Boston: Beacon Press, 1994); a pioneering book—Nabhan is a field biologist, naturalist, and writer, and Trimble is a nature photographer and naturalist.

3. E. A. McIlhenny, *The Alligator's Life History* (Boston: Christopher Publishing House, 1935 [reprinted by the Society for the Study of Reptiles and Amphibians, 1976]), 7, 8. The extremely rare first edition in my library is one of my most prized possessions.

4. J. R. McNeill, *Something New under the Sun: An Environmental History of the Twentieth-Century World* (New York: Norton, 2001).

5. The magnificent field guides that are now widely available have shown a continuous improvement over the original, excellent *Guide to the Birds*, first published by the artist and naturalist Roger Tory Peterson in 1934. The first printing of this book, two thousand copies, sold out in one week, and it has gone through a number of editions since. It was followed by an explosion of affordable field guides in every conceivable area of natural history, from stars and planets, to rocks and minerals, to fungi, seashore life, flowering plants, trees, insects, fish, reptiles and amphibians, birds, mammals, animal tracks, and even protozoa and bacteria. The Peterson Field Guide series, published by Houghton Mifflin, covers quite a few of these areas, but many other publishers have also provided fine, beautifully illustrated books. Although they are too numerous to list here, the following are a few that I use—the two bird guides reflect my eastern bias, but western editions are also available: Lawrence Newcomb, *Newcomb's Wildflower Guide* (New York: Little, Brown, 1977); Roger Tory Peterson and Virginia M. Peterson, *Birds of Eastern and Central North America*, 5th ed. (Boston: Houghton Mifflin, 2002); David Allen Sibley, *The Sibley Field Guide to Birds of Eastern*

*North America* (New York: Knopf, 2003); Olaus J. Murie and Mark Elbroch, *A Field Guide to Animal Tracks*, 3rd ed. (Peterson Field Guide Series) (Boston: Houghton Mifflin, 2005).

The best Web site for bird-watchers, beginners to advanced, is that of the Cornell Lab of Ornithology/Audubon Society at http://www.ebird .org. This highly interactive site, supported by the National Science Foundation, includes information on every aspect of North American birds— where to find them, their habits, their population trends, rare sightings, local bird-watching groups from Mexico to Canada to Alaska, choosing binoculars, and so on. Browsing this attractive and well-written site is almost as much fun as looking at the birds themselves. Relying heavily on carefully collected data from its thousands of bird-watchers, this site is a beautiful example of science generated by the public, known as "citizen science." The Cornell Lab's large staff of scientists ensures the accuracy of the information presented by ebird.

Many of the best books on natural history are older books, harking back to the days when it was still a popular subject in college and high school. The following are examples of the mostly older works: Tom Brown Jr., *Tom Brown's Field Guide to the Forgotten Wilderness* (New York: Berkley, 1987) (The "wilderness" is in and around your city or suburb, and Tom Brown is America's premier teacher of wilderness awareness and tracking. A fascinating guide to the daily enjoyment of nature—anywhere.); Anna Botsford Comstock, *Handbook of Nature Study* (Ithaca, NY: Cornell University Press [1939] 1986) (This book contains 887 pages of priceless, clearly written information about plants, animals, minerals, soil, weather, and the skies above.); Steven D. Garber, *The Urban Naturalist* (New York: John Wiley, 1987); John Kieran, *A Natural History of New York City* (Boston: Houghton Mifflin, 1959); Aldo Leopold, *A Sand County Almanac and Sketches Here and There* (New York: Oxford University Press, 1949); May Theilgaard Watts, *Reading the Landscape of America* (New York: Collier Macmillan, 1975). (Watts describes the combination of detective ability and ecological knowledge required to understand the past and predict the future of the landscapes around us; see especially chapter 7.) R. S. R. Fitter, *London's Natural History* (London: Collins, 1945). (Nature in and around London, from Neolithic times until the mid-twentieth century.)

## Teaching Field Ecology

1. See the discussion of field guides and other information sources on natural history given in note 5 of the chapter titled "Strangers in Our Own Land."

## The Ubiquitous Right-of-Way

1. Frank E. Egler and Stan R. Foote, *The Plight of the Rightofway Domain: Victim of Vandalism* (Mt. Kisco, NY: Futura Media Services, 1975).
2. James Howard Kunstler, *The Geography of Nowhere: The Rise and Decline of America's Man-Made Landscape* (New York: Simon and Schuster, 1993). Not specifically about rights-of-way (few popular books are), this book explains how the automobile and its associated suburbs have altered the geography of America.
3. Rob Edwards and Duncan Graham-Rowe, "Guilty as Charged: Powerful Fields From Pylons and Cables Are Linked to Childhood Cancer," *New Scientist* (Mar. 10, 2001): 7.
4. William A. Niering and Richard H. Goodwin, "Creation of Relatively Stable Shrublands with Herbicides: Arresting 'Succession' on Rights-of-Way and Pastureland," *Ecology* 55 (1974): 784–795.

## A Walk in the Woods

1. Oliver Rackham, *The History of the Countryside* (London: Phoenix (Orion Books Ltd.) [1986] 1997). Also see: W. G. Hoskins, *The Making of the English Landscape* (Harmondsworth, U.K.: Penguin Books, 1992); John Perlin, *A Forest Journey: The Role of Wood in the Development of Civilization* (Cambridge, MA: Harvard University Press, 1991).
2. John E. Kuser, ed., *Urban and Community Forestry in the Northeast*, 2nd ed. (New York: Springer, 2006).
3. American forests also have their history. See: William Cronon, *Changes in the Land: Indians, Colonists, and the Ecology of New England* (New York: Hill and Wang, 1983); Tom Wessels, *Reading the Forested Landscape: A Natural History of New England* (Woodstock, VT: Countryman Press, 1997); Leslie Jones Sauer and Andropogon Associates, *The Once and Future Forest: A Guide to Forest Restoration Strategies* (Washington, DC: Island Press, 1998)—an exciting book that describes what the temperate forests of the eastern United States

once were, and what they might, with wise restoration, become again, not only in rural areas but in the suburbs and cities as well.

## Old Growth

1.  Howard Berkes, "Pentagon Memorial," http://www.npr.org/templates/ story/story.php?storyId=1132931 (accessed April 30, 2008).

2.  Serge Payette and Ann Delwaide, "Dynamics of Subarctic Wetland Forests over the Past 1500 Years," *Ecological Monographs* 74 (Aug. 2004): 373–391; Lily Wang, Serge Payette, and Yves Bégin, "1300-Year Tree-Ring Width and Density Series Based on Living, Dead and Subfossil Black Spruce at Tree-Line in Subarctic Quebec, Canada," *The Holocene* 11 (2001): 333–341; Dominique Arseneault and Serge Payette, "Reconstruction of Millennial Forest Dynamics From Tree Remains in a Subarctic Tree Line Peatland," *Ecology* 78 (1997): 1873–1883.

3.  Claude Morneau and Serge Payette, "Long-Term Fluctuations of a Caribou Population Revealed by Tree-Ring Data," *Canadian Journal of Zoology* 78 (2001): 1784–1790.

4.  International Society for Ecology and Culture, *Ancient Futures: Learning from Ladahk* (Berkeley, CA: ISEC, 1993), VHS, based on the book by Helena Norberg-Hodge, *Ancient Futures: Learning from Ladakh* (San Francisco: Sierra Club Books, 1991).

5.  Leopold Kohr, *The Breakdown of Nations* (New York: Dutton, [1957] 1978), xviii; E. F. Schumacher, *Small Is Beautiful: Economics as if People Mattered* (Point Roberts, WA: Hartley and Marks, [1973], 1995).

## Intimacy with Nature

1.  Edwin Way Teale, ed., *The Wilderness World of John Muir* (Boston, Houghton Mifflin, 1954), 190; contains some of the best of Muir's writing, including many extracts that portray his fearless intimacy with nature. Robert Finch and John Elder, eds., *The Norton Book of Nature Writing*, college ed. (New York: Norton, 2002). *The Norton Book of Nature Writing* is an enormously readable and comprehensive anthology of nature writing covering a wide range of human degrees of intimacy with nature over the past two centuries. The selections are by authors as varied as Gilbert White, John Bartram, Meriwether Lewis, Charles Darwin, Mark Twain, Isak Dinesen,

John Fowles, Annie Dillard, Barbara Kingsolver, and my own teacher, Archie Carr. Peter Sauer, ed., *Finding Home: Writing on Nature and Culture from Orion Magazine* (Boston: Beacon Press, 1992). The essays in *Finding Home* explore many of the exciting ways that people can develop an intimacy with nature.

## Part Five: Restoring the Community

1. Wendell Berry, *Life Is a Miracle: An Essay Against Modern Superstition* (Washington, DC: Counterpoint, 2000), 150.

## The Utopia Fallacy

1. *Pirkei Avot (The Ethics of the Fathers)*, a first to third century rabbinic commentary available in many editions; also a common Arabic proverb.
2. John Gray, *False Dawn: The Delusions of Global Capitalism* (London: Granta Books, 1998); International Forum on Globalization, *Alternatives to Economic Globalization* (San Francisco: Berrett-Koehler, 2002); Jerry Mander and Edward Goldsmith, eds., *The Case against the Global Economy: And for a Turn Toward the Local* (San Francisco: Sierra Club Books, 1996); James Goldsmith, *The Trap* (New York: Carroll and Graf, 1994); George Soros, *Soros on Soros* (New York: John Wiley and Sons, 1995).
3. David Ehrenfeld, "The Environmental Limits to Globalization," *Conservation Biology* 19 (April 2005): 321. See also: James Howard Kunstler, *The Long Emergency: Surviving the End of Oil, Climate Change, and Other Converging Catastrophes of the Twenty-First Century* (New York: Grove Press, 2006) (This is an honest, informed look at our energy problems and the technological alternatives that have been suggested.); David Ehrenfeld, "Recalculating Future Oil Reserves," *Science* 309 (2005): 54–55.
4. Vaclav Smil, *Energy at the Crossroads: Global Perspectives and Uncertainties* (Cambridge, MA: MIT Press, 2003), 371–372.
5. John Ralston Saul, *Voltaire's Bastards: The Dictatorship of Reason in the West* (New York: Vintage, 1992), 31.
6. J. R. R. Tolkien, *The Lord of the Rings*, vol. 3, *The Return of the King* (London: George Allen and Unwin, 1954), 310. Tolkien's brand of realistic, non-utopian hope, which I share, is beautifully and movingly described by the British scholar Patrick Curry. Note especially his last chapter: "Conclusion:

Hope without Guarantees," in *Defending Middle-Earth: Tolkien, Myth and Modernity* (New York: St. Martin's Press, 1997).

7. Aleksandr I. Solzhenitsyn, "To Tame Savage Capitalism," *New York Times*, Nov. 28, 1993, E11.

8. As I write, early signs of this process are beginning to appear. For example, the reckless expansion of the credit market for housing construction (subprime loans), combined no doubt with the rapidly increasing cost of energy and building materials, is resulting in whole neighborhoods of unsalable new houses boarded up with plywood: David Gonzalez, "Risky Loans Help Build Ghost Town of New Homes," *New York Times*, Sept. 24, 2007, B1, B5.

9. See the chapter titled "Pretending," p. 3.

10. Jared Diamond, *Guns, Germs, and Steel: The Fates of Human Societies* (New York: Norton, 1997), 54–57.

11. Tim Hunkin, "Just Give Me the Fax," *New Scientist* 137 (Feb. 13, 1993): 33–37.

12. Richard Heinberg, *Fifty Million Farmers*, Twenty-Sixth Annual E. F. Schumacher Society Lectures (Great Barrington, MA: E. F. Schumacher Society, 2006). A fascinating account of Cuba's switch from industrial to ecological agriculture, which has increased food production and availability while greatly lowering fuel dependence and pesticide and fertilizer costs.

13. E. F. Schumacher, *Small Is Beautiful: Economics as if People Mattered: 25 Years Later . . . With Commentaries* (Point Roberts, WA: Hartley and Marks, 1999), 46. This edition of the classic work, published on the twenty-fifth anniversary of its first appearance, contains helpful updates by more than thirty experts in economics, energy and technology, community development, ecological sustainability, and soils. It also has an appendix on resources available in these areas.

14. Martin Van Creveld, *The Transformation of War* (New York: Free Press, 1991). In this book, an influential military historian and analyst predicted the nature of the likely military conflicts of the twenty-first century. The Israeli-Palestinian and Iraq conflicts have tragically confirmed Van Creveld's predictions about the nature of modern warfare.

15. Isaiah Berlin, "The Pursuit of the Ideal," in Isaiah Berlin, *The Proper Study of Mankind: An Anthology of Essays*, ed. Henry Hardy and Roger Hausheer (New York: Farrar, Straus and Giroux, 1998), 15.

16. Hildegarde Hannum, ed., *People, Land, and Community: Collected E. F. Schumacher Society Lectures* (New Haven: Yale University Press, 1997). The essays in this collection contain many practical, nonutopian ideas for preparing now for the days ahead: from local currencies to the Garden Project. The

lectures since 1997 are available as separate pamphlets from the E. F. Schumacher Society, Great Barrington, MA.

## Traditions

1. *Pentateuch and Haftorahs*, 2nd ed., ed. J. H. Hertz (London: Soncino, 1978), 525.
2. Arthur Schaffer, "The Agricultural and Ecological Symbolism of the Four Species of Sukkot," in *Judaism and Environmenal Ethics*, ed. Martin Yaffe (Lanham, MD: Lexington, 2001), 112–124. Although the Schaffer article was originally published in *Tradition* 20 (1982), this version may be more accessible to the reader. Much of the discussion of Sukkot in this chapter is based on this article, which is also the source of the relevant quotations.
3. Eric Hobsbawm and Terrence Ranger, eds., *The Invention of Tradition* (Cambridge: Cambridge University Press, 1983), 13.
4. Maulana Karenga, *Kwanzaa: A Celebration of Family, Community and Culture* (Los Angeles: University of Sankore Press, 1998). Also see http://www .officialkwanzaawebsite.org (accessed Aug. 1, 2007).
5. The following four books, although not confined to the subject of traditions, contain much useful material on the interaction between religious traditions and ecology: Seyyed Hossein Nasr, *Religion and the Order of Nature* (New York: Oxford University Press, 1996); Ellen Bernstein, ed., *Ecology and the Jewish Spirit: Where Nature and the Sacred Meet* (Woodstock, VT: Jewish Lights Publishing, 1998); John E. Carroll, Paul Brockelman, and Mary Westfall, eds., *The Greening of Faith: God, the Environment, and the Good Life* (Hanover, NH: University Press of New England, 1997); Mawil Izzi Dien, *The Environmental Dimensions of Islam* (Cambridge, U.K.: Lutterworth Press, 2000).

## Jane Austen and the World of the Community

1. Deirdre Le Faye, *Jane Austen's Letters*, 3rd ed. (New York: Oxford University Press, 1995), 324, 216, 57–58, 312. Any spelling irregularities in quotes from Austen's letters reflect her own orthographic preferences.
2. Tony Tanner, *Jane Austen* (Cambridge: Harvard University Press, 1986), 102.
3. There are numerous editions of Jane Austen's six major novels, many with excellent critical commentaries. The classic edited set, and perhaps the most attractive is *The Oxford Illustrated Jane Austen*, 3rd ed., ed. R. W. Chapman and M. Lascelles (Oxford: Oxford University Press, 1988).

4. Biographies and critical studies of Jane Austen abound; the excellent one by Tanner is cited in note 2. To understand Jane Austen as a real person and a member of a set of related, vibrant communities—to encounter her fully in her age and place—I recommend the magnificent biography by Claire Tomalin: *Jane Austen: A Life* (New York: Vintage Books, [1997] 1999). Once started, it is hard to put down.

## Universities, Schools, and Communities

1. See the notes for the chapter "I Reinvent Agriculture."
2. Jonathan Kozol, *Amazing Grace: The Lives of Children and the Conscience of a Nation* (New York: HarperCollins, 1996). Similarly, only a few city blocks from many well-equipped university classrooms, with comfortable seats and wireless access to the Internet, are dangerous city schools that lack the bare necessities for learning. See also Jonathan Kozol, *Letters to a Young Teacher* (New York: Crown, 2007).
3. Muhammad Yunus, with Alan Jolis, *Banker to the Poor: Micro-Lending and the Battle against World Poverty* (New York: Public Affairs, 1999), 89. This is one of the most hopeful and practical books cited in these pages. Yunus shows that it is possible to change the world for the better on a grand scale. In 2006, he and his Grameen Bank shared the Nobel Peace Prize.
4. One distinguished professor in a genetics department of a large, midwestern university told me that he is expected to bring in each year at least $100 in grant money overhead for each square foot of laboratory space he occupies—anything less, and his support from the university is diminished. If he wants to pursue the larger issues that affect society—but which don't pay immediate returns in cash—he has to divert other research funds for the purpose or take the money from his own pocket (he does).
5. Chargaff was cited in the chapter titled "Accelerating Social Evolution": Erwin Chargaff, *Heraclitean Fire: Sketches from a Life before Nature* (New York: Rockefeller University Press, 1978), 161.
6. The following books describe the plight of the modern university—cut off from community, moral purpose, and the natural world—and provide a vision of the direction that education must follow if the best parts of our culture and environment are to survive in the decades ahead: Bruce Wilshire, *The Moral Collapse of the University: Professionalism, Purity, and Alienation* (Albany: State University of New York Press, 1990); Page Smith, *Killing the Spirit: Higher Education in America* (New York: Viking Penguin, 1990); Stan

Rowe, "Role of the University," in *Home Place: Essays on Ecology* (Edmonton, Alberta: NeWest, 1990), 129–138; David W. Orr, *Earth in Mind: On Education, Environment, and the Human Prospect* (Washington, DC: Island Press, 1994).

## What Do We Owe Our Children?

1. Benjamin Spock and Robert Needlman, *Dr. Spock's Baby and Child Care*, 8th ed. (New York: Pocket Books, 2004). A later and thicker edition than the one I used—it is good to see that this invaluable book is being revised and kept in print. There is now a wide assortment of good books on baby care.
2. *The Babylonian Talmud*, tractate Kiddushin, 29a, 30b.
3. Wendell Berry, *The Unsettling of America: Culture and Agriculture* (San Francisco: Sierra Club Books, 1977), 23, 26.

## Epilogue: A Call for Fusion and Regeneration

1. G. W. F. Hegel, Preface to *Elements of the Philosophy of Right*, 1820. http://www.marxists.org/reference/archive/hegel/works/pr/preface.htm (accessed April 25, 2008).
2. Edmund Morris, *Theodore Rex* (New York: Modern Library, 2001), 500.
3. Nathan Miller, *Theodore Roosevelt: A Life* (New York: William Morrow, 1992), 337.
4. Curt Meine and Richard L. Knight, eds., *The Essential Aldo Leopold: Quotes and Commentaries* (Madison: University of Wisconsin Press, 1999), 184.
5. Charles Van Hise, *The Conservation of Natural Resources in the United States* (New York: Macmillan, [1910] 1916), 377.
6. David Ehrenfeld, *Beginning Again: People and Nature in the New Millennium* (New York: Oxford University Press, [1993], 1995), 65–72. This chapter, "Forgetting," cited earlier, explores the problem of loss of knowledge in more detail than I do here.
7. The reasons why the present global system is so unstable, and why a more local, less centralized world may be coming, have been carefully set forth by the archaeologist Joseph Tainter. The economic and historical arguments he makes are both convincing and frightening: Joseph A. Tainter, *The Collapse of Complex Societies* (Cambridge: Cambridge University Press, 1988).

8. Kwame Anthony Appiah, *Cosmopolitanism: Ethics in a World of Strangers* (New York: Norton, 2006), xx.

9. William J. Mitsch and James G. Gosselink, *Wetlands*, 3rd ed. (New York: John Wiley and Sons, 2000). See especially: chap. 16, "Values and Valuation of Wetlands"; chap. 17, "Human Impacts and Management of Wetlands"; chap. 19, "Wetland Creation and Restoration"; and chap. 20, "Treatment Wetlands."

10. Wes Jackson, *New Roots for Agriculture* (Lincoln: University of Nebraska Press, 1985); Wes Jackson, "The Agrarian Mind: Mere Nostalgia or a Practical Necessity?" in *The Essential Agrarian Reader: The Future of Culture, Community, and the Land*, ed. Norman Wirzba (Lexington: The University Press of Kentucky, 2003), 140–153; Gary Paul Nabhan, *The Desert Smells Like Rain: A Naturalist in Papago Indian Country* (San Francisco: North Point Press, 1982); Gary Paul Nabhan, *Why Some Like It Hot: Food, Genes, and Cultural Diversity* (Washington, DC: Island Press, 2004).

11. Carlos F. Amábile-Cuevas, "New Antibiotics and New Resistance," *American Scientist* 91 (March–April 2003): 138–149. The subtitle to this article reads: "In many ways, the fight against antibiotic resistance is already lost; preventing bacterial disease requires thoughtful new approaches." Richard Stone, "Stalin's Forgotten Cure," *Science* 298 (2002): 728–731. The older, Russian idea of treating antibiotic-resistant infections with bacteriophages, viruses that eat bacteria, is an example of a novel, ecologically inspired way of treating infectious diseases. See also the discussion of antibiotic resistance in the chapter "The Downside of Corporate Immortality."

12. www.quiviracoalition.org (accessed August 30, 2007).

13. Hazel Henderson, Marginal comment in E. F. Schumacher, *Small Is Beautiful: Economics as if People Mattered: 25 Years Later . . . With Commentaries* (Point Roberts, WA: Hartley and Marks, 1999), 40–41.

14. Information about every phase of the BerkShare story can be obtained from the Web site, www.berkshare.org, and from the E. F. Schumacher Society Web site, www.smallisbeautiful.org (both sites accessed August 30, 2007).

15. www.freecycle.org (accessed August 30, 2007).

16. For some of the most creative and informed (if sometimes controversial) discussions of how life is going to change in North America, and what we can do to steer the change in a benign and livable direction, see the Web site of Julian Darley's Post Carbon Institute, www.postcarbon.org, and James Howard Kunstler's home page, www.Kunstler.com (both accessed May 19, 2008).

# CREDITS

Epigraphs in *Becoming Good Ancestors* have been taken from the following sources:

Two selections from *Life Is a Miracle*, by Wendell Berry, copyright © 2000 by Wendell Berry. Used by permission of Counterpoint, a member of Perseus Books Group.

One selection from *The Collected Poems of Thomas Merton*, by Thomas Merton, *extract from "A Letter to Pablo Antonio Cuadra Concerning Giants,"* copyright © 1963 by The Abbey of Gethsemani, Inc. Reprinted by permission of New Directions Publishing Corp. and Laurence Pollinger Limited.

Two selections from *Voltaire's Bastards: The Dictatorship of Reason in the West*, by John Ralston Saul, copyright © 1992 by John Ralston Saul. Used by permission of Vintage Books, a division of Random House, Inc.

One selection from *Building a Bridge to the 18th Century: How the Past Can Improve Our Future*, by Neil Postman, copyright © 1999 by Neil Postman. Used by permission of Vintage Books, a division of Random House, Inc.

# INDEX

CPSIA information can be obtained at www.ICGtesting.com
Printed in the USA
BVOW080830091012

302542BV00001B/69/P